WAVES

A Macmillan Physics Text

Consulting Editor: Professor P. A. Matthews, F.R.S.

Other titles
MODERN ATOMIC PHYSICS: FUNDAMENTAL PRINCIPLES: *B. Cagnac and J.-C. Pebay-Peyroula*
MODERN ATOMIC PHYSICS: QUANTUM THEORY AND ITS APPLICATIONS: *B. Cagnac and J.-C. Pebay-Peyroula*

Nature–Macmillan Series
AN INTRODUCTION TO SOLID STATE PHYSICS AND ITS APPLICATIONS: *R. J. Elliott and A. F. Gibson*
THE SPECIAL THEORY OF RELATIVITY: *H. Muirhead*
FORCES AND PARTICLES: *A. B. Pippard*

Forthcoming
AN INTRODUCTION TO NUCLEAR PHYSICS: *D. M. Brink, G. R. Bishop and G. R. Satchler*
SYMMETRY IN PHYSICS: *J. P. Elliott and P. G. Dawber*
ELECTRICITY AND MAGNETISM: *B. L. Morgan and R. W. Smith*

WAVES

D. R. Tilley

University of Essex

MACMILLAN

First published 1974 by
THE MACMILLAN PRESS LTD
London and Basingstoke
Associated companies in New York Dublin
Melbourne Johannesburg and Madras

SBN 333 15464 9 (hard cover)
333 16612 4 (paper cover)

Printed in Great Britain
WILLIAM CLOWES & SONS, LIMITED
London, Colchester and Beccles

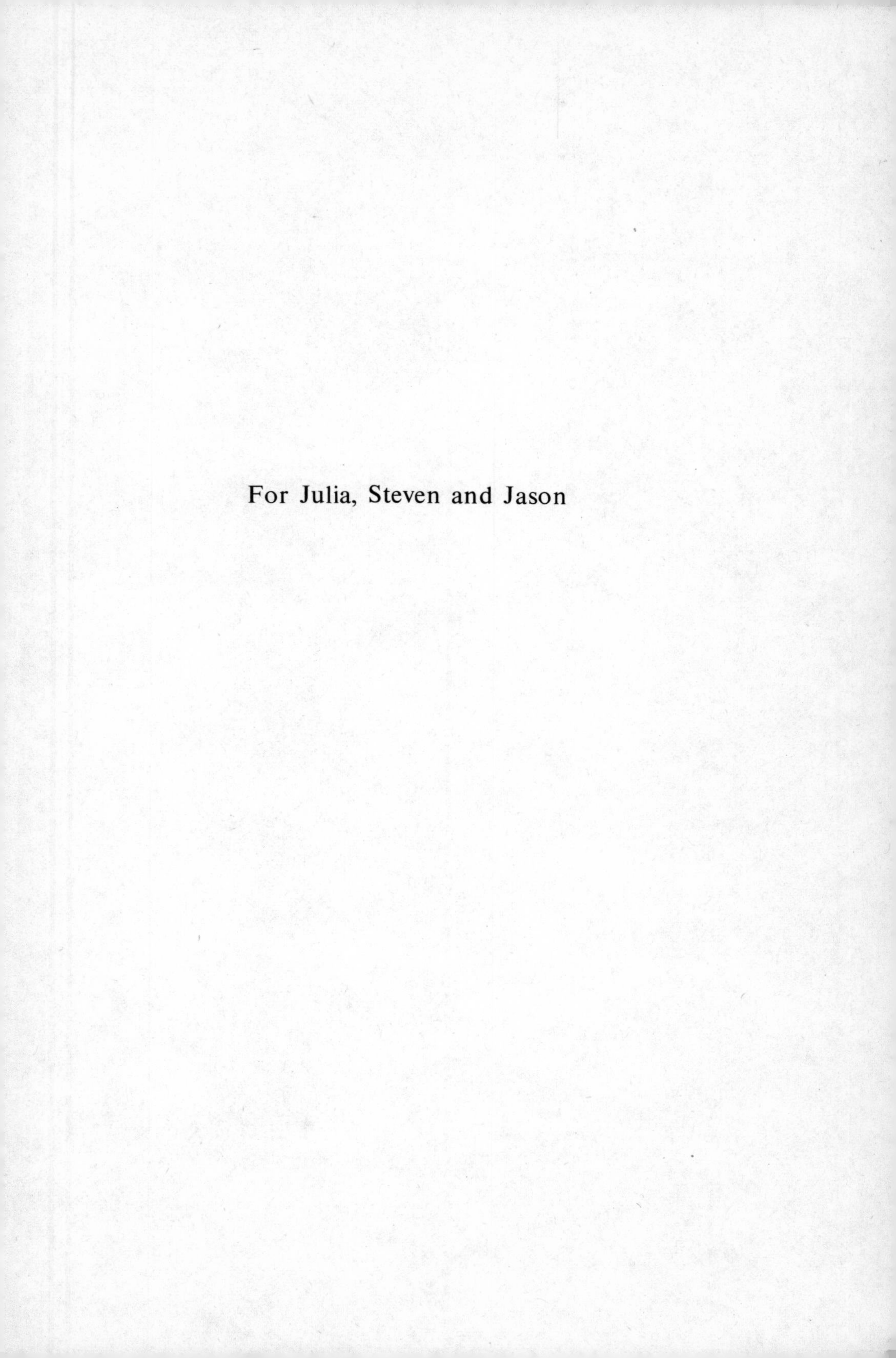

For Julia, Steven and Jason

Contents

Preface

This book gives an introductory account of the ideas about wave motion and resonance which are important in modern physics, and is based on a course of lectures I have given at the University of Essex for a number of years.

I have tried to emphasise those concepts which are most frequently used at a more advanced level. In particular I have stressed the importance of relating a waveform in time to its frequency amplitude. It is not possible in an introductory text to give a completely mathematical account using Fourier transform theory, but the subject is so important that an early qualitative introduction is called for. The wave equation is given relatively less emphasis than has become usual, and an account of it is deferred until chapter 6. This re-ordering enables one to concentrate on what I feel are more essential matters and to deal with them in a fairly discursive manner. I have not confined the subject matter to classical waves, since the sections on wave–particle duality and the Schrödinger equation have a natural place in a book of this kind.

I have assumed that the reader has a prior knowledge of elementary calculus and trigonometry, and is familiar with the simple harmonic oscillator and a few elementary properties of wave motion. Most students using this book will attend a parallel mathematics course, and I have therefore allowed the mathematical level to drift upwards as necessary in the last two chapters. The book is obviously not a specialist work on optics, acoustics or wave mechanics. However, it should provide a suitable introduction to the courses which a specialist will meet later, and at the same time offer a worthwhile excursion in physics for the student who will later specialise in some other discipline.

To illustrate certain points, a number of worked examples have been included in the text. In addition, there are problems at the end of each chapter, those marked UE being taken from the University of Essex first-year examinations.

I am grateful to many of my colleagues and to my students for their help in contributing ideas and in removing errors. I should like to thank Mr E. Adair for photographic assistance, and Mrs M. Baker for her very capable typing of the manuscript.

D. R. TILLEY

1

The Simple Harmonic Oscillator and Superposition of Vibrations

In this book, we intend to discuss general features of the propagation of waves. We shall define a wave rather broadly as anything that produces a displacement $u(x, t)$ where x is a spatial co-ordinate and t is time. In the simple example of a wave on a stretched string, we may take u as the sideways, or transverse displacement of the string from the straight line that is its undisturbed position. Again, in a surface wave on water, u is the vertical displacement of the surface from a plane. In both these examples, we are dealing with the displacement of a medium,

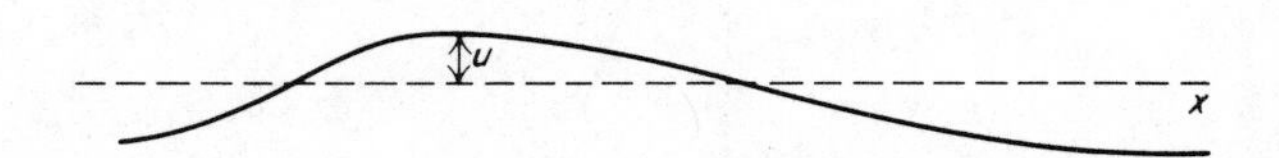

Figure 1.1 On a stretched string, $u(x, t)$ is the transverse displacement

either the string or the water. In the important case of light travelling through a vacuum, however, we must interpret 'displacement' differently. It will be recalled that light is not propagated by means of vibrations of a jelly-like aether; we must simply regard the displacement as an oscillating electric field, with which is associated an oscillating magnetic field.

We shall find that it is often convenient to treat a general wave as a superposition of waves of a single frequency, since the latter generally travel through the medium in a simpler way than a general wave form. In a single-frequency wave, the displacement at any point x is a sinusoidal function of t; that is, it varies with time like the displacement from equilibrium of a simple harmonic oscillator. An

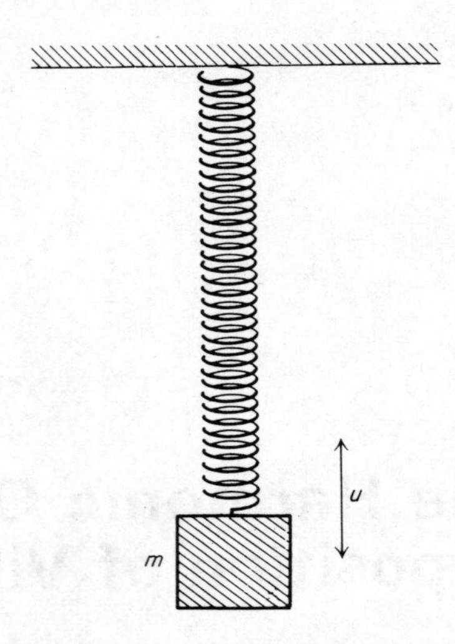

Figure 1.2 A mass on a spring. u is the vertical displacement from the equilibrium position, m is the mass and K is determined by the spring constant

understanding of the simple harmonic oscillator is therefore basic to a study of wave motion. We shall see further (chapter 5) that the harmonic oscillator serves as a very useful model for the response of a medium to an incident wave. For these reasons, this chapter is devoted to the simple harmonic oscillator. We briefly

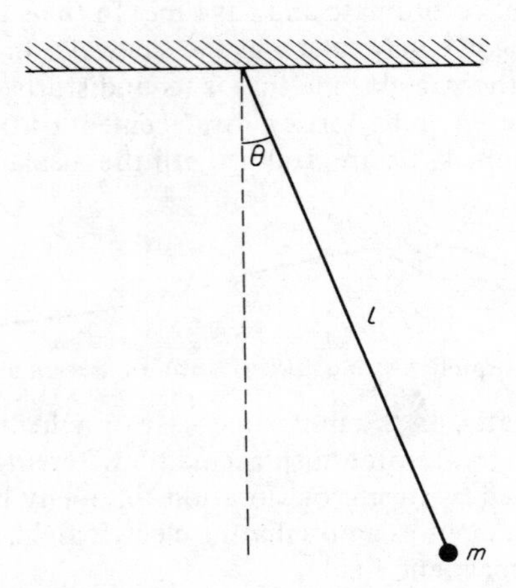

Figure 1.3 For small angular displacements θ a simple pendulum undergoes simple harmonic motion. u is the angle θ, m is the mass and K is g/l

review the defining equation and its solutions in section 1.1, and then in sections 1.2 and 1.3 we discuss the superposition of harmonic vibrations to produce a displacement that is a general function of time. The ideas introduced in these sections will recur throughout the discussion of wave motion.

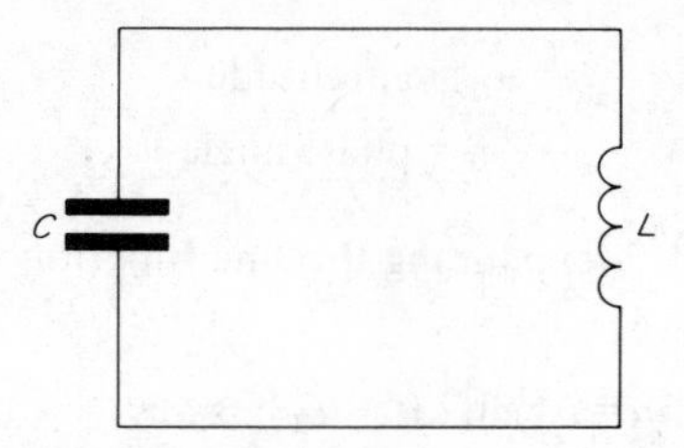

Figure 1.4 A 'ringing circuit' composed of a capacitor and an inductor is an example of simple harmonic motion. u may be taken as the charge Q on either condensor plate, m is replaced by L and K by $1/C$

1.1 Simple Harmonic Oscillator

A simple harmonic oscillator is any system that moves under a restoring force proportional to the displacement from an equilibrium point; some examples are shown in figures 1.2 to 1.4. In all those cases the equation of motion is

$$m\frac{d^2u}{dt^2} = -Ku \tag{1.1}$$

We take as known the general solution of this equation, namely

$$u = u_0 \sin(\omega t + \eta) \tag{1.2}$$

with

$$\omega^2 = K/m \tag{1.3}$$

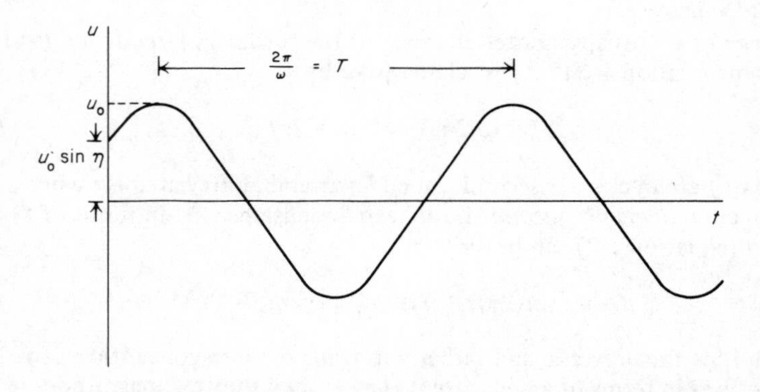

Figure 1.5 General solution of harmonic oscillator equation

This solution is sketched in figure 1.5. (If the reader is not familiar with the solution of the harmonic oscillator equation, he will find a treatment in any A-level applied mathematics text.) We note that equation 1.1 is second order in t (that is, it involves $\mathrm{d}^2u/\mathrm{d}t^2$) and consequently the solution involves two constants of integration. As we have written the solution, these are

$$u_0 = \text{amplitude}$$
$$\eta = \text{phase angle}$$

We can rewrite equation 1.2, expanding the sine function with the appropriate identity from appendix 1, as

$$u = u_1 \sin \omega t + u_2 \cos \omega t \tag{1.4}$$

with

$$u_1 = u_0 \cos \eta$$
$$u_2 = u_0 \sin \eta$$

so that now the constants of integration appear as u_1 and u_2. The constants of integration, u_0 and η, or equivalently u_1 and u_2, may be determined by initial conditions on the motion of the particle, for example the initial values of position and velocity.

We have written our solution in terms of the *angular frequency* ω. We can easily relate ω to the *periodic time* T for the motion to repeat itself; the system returns to its initial displacement and velocity when the argument of the sine function in equation 1.2 has increased by 2π so we have

$$\omega T = 2\pi \tag{1.5}$$

That is, the periodic time is given by

$$T = 2\pi/\omega \tag{1.6}$$

as marked in figure 1.5.

Frequencies are usually quoted in terms of the (ordinary) *frequency* $f = 1/T$; we see from equation 1.6 that f is related to ω by

$$f = \omega/2\pi \quad \text{or} \quad \omega = 2\pi f \tag{1.7}$$

The units of f are cycles per second, called *hertz* and abbreviated Hz whereas the units of ω are *radians per second.* Both have dimensions s^{-1}. In terms of f our oscillation (equation 1.2) can be written

$$u = u_0 \sin(2\pi ft + \eta) = u_0 \sin(2\pi t/T + \eta) \tag{1.8}$$

We shall follow the universal and rather confusing modern convention of writing our oscillations in terms of angular frequency ω, and quoting magnitudes in hertz which are units of f.

Worked example 1.1

A simple harmonic oscillator of frequency $\omega = 2\pi\ \mathrm{s}^{-1}$ is set into motion with the initial conditions

$$u = 5\sqrt{3}$$

$$\frac{\mathrm{d}u}{\mathrm{d}t} = -10\pi$$

Find the equation of the subsequent motion, expressing it both in the 'sine plus cos' form of equation 1.4, and in the 'amplitude-phase' form of equation 1.2.

Answer

Let

$$u = u_1 \sin \omega t + u_2 \cos \omega t$$

Then

$$\frac{\mathrm{d}u}{\mathrm{d}t} = u_1 \omega \cos \omega t - u_2 \omega \sin \omega t$$

At $t = 0$, therefore

$$u = u_2$$

$$\frac{\mathrm{d}u}{\mathrm{d}t} = u_1 \omega$$

Putting in the numerical values given, we find

$$u_2 = 5\sqrt{3}$$

$$u_1 = -5$$

and the motion in 'sine plus cos' form is

$$u = 5\sqrt{3} \cos \omega t - 5 \sin \omega t$$

with $\omega = 2\pi$. To get the 'amplitude-phase' form, we take out the factor $(u_1^2 + u_2^2)^{1/2} = 10$. This gives

$$\begin{aligned} u &= 10\left(\frac{\sqrt{3}}{2} \cos \omega t - \tfrac{1}{2} \sin \omega t\right) \\ &= 10(\cos 30^\circ \cos \omega t - \sin 30^\circ \sin \omega t) \\ &= 10 \cos(\omega t + 30^\circ) \end{aligned}$$

which is the required form.

1.2 Superposition of Vibrations—Equal Frequencies

As we have already pointed out, we shall frequently have to deal with superposition of vibrations or wave motions. The simplest instance is the superposition of harmonic vibrations which all have the same frequency; this is basic to all *interference* problems. By way of example, consider the Young's slits experiment shown in figure 1.6. The experiment is designed to observe the interference between the trains of light coming from S_1 and S_2, which are narrow transparent slits in an opaque screen. Our sketch is a section through the experimental set up; the slits extend a certain distance in the direction perpendicular to the plane of the sketch. S_1 and S_2 are illuminated by light from the lamp G, which is a source of

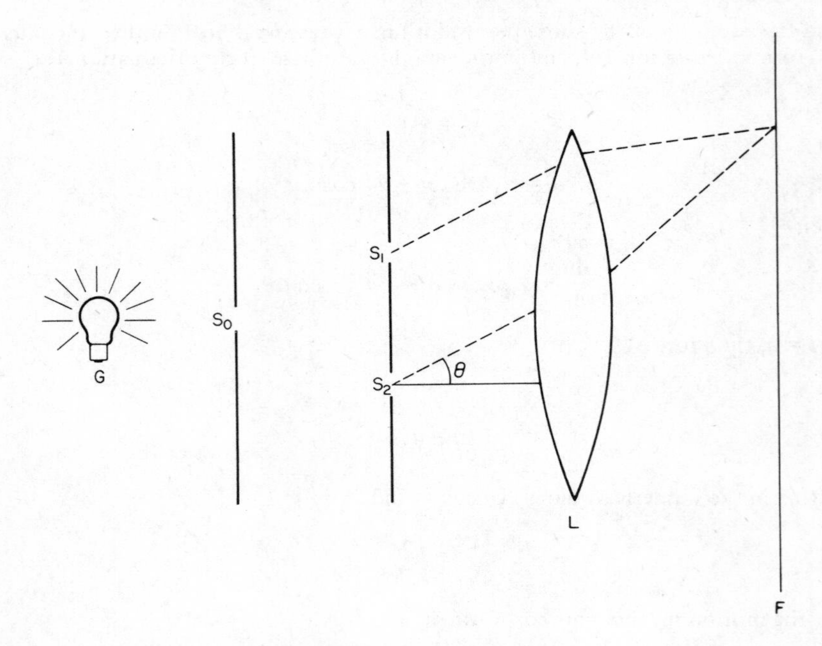

Figure 1.6 Young's slits experiment

monochromatic (single-frequency) light, say a gas discharge lamp of some kind. The light from G first passes through the slit S_0, which has to be a very narrow slit, for reasons to which we shall return at the end of this section. It is simplest to study the variation with angle θ of the transmitted light, so that a lens L is placed after S_1 and S_2, and the light intensity observed in the focal plane F of L. With the arrangement described, the pattern of intensity observed is a series of light and dark *fringes,* as shown in figure 1.7.

The fringe pattern has its origin in the different distances travelled by the beams of light coming from the two slits. As shown in figure 1.8, the difference in path length is $d \sin \theta$, where d is the distance between the slits. If $d \sin \theta$ is exactly equal to the wavelength λ, then the crests of the S_2 wave fall on the crests of the S_1 wave, and the waves reinforce one another. We can say equivalently that the phase difference between the waves is 2π. Similarly, if $d \sin \theta$ is equal to $\lambda/2$, the phase difference is π and the waves cancel one another. For a general

Figure 1.7 Fringes observed in Young's slits experiment. The fall off in intensity at wide angles is due to the finite width of the slits. (Reproduced with permission from M. Cagnet *et al., Atlas of Optical Phenomena,* Springer, Berlin (1962))

value of $d \sin\theta$, the phase difference is $2\pi(d \sin\theta/\lambda)$, and we may write the complete wave as

$$u = u_0 \sin(\omega t + \eta) + u_0 \sin(\omega t + \eta + \delta) \tag{1.9}$$

with

$$\delta = 2\pi(d \sin\theta)/\lambda \tag{1.10}$$

Here u_0 is the amplitude of the wave which would come through either slit in isolation, and η is the initial phase of the S_1 wave; it has the same meaning as η in equation 1.2 and figure 1.5.

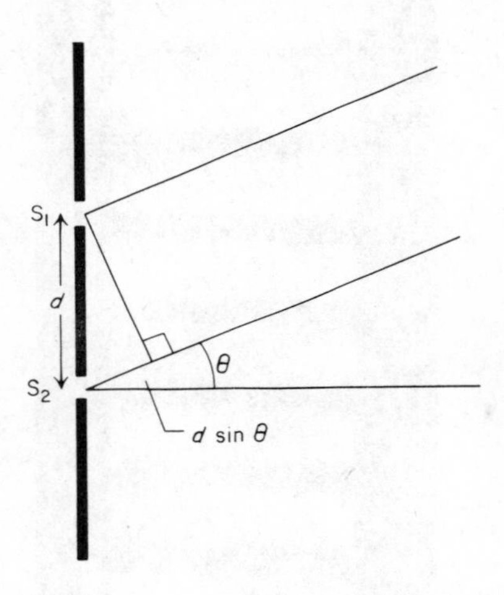

Figure 1.8 Path difference in Young's slits experiment

In writing down equation 1.9 we have made some assumptions which we should mention explicitly. First, the slits themselves have a finite width, say d_s, and this implies some phase variation between the wave trains emitted from different parts of the same slit. We shall see in section 3.2 that this leads to significant changes in the fringes only on the angular scale $\sin\theta_s \approx \lambda/d_s$, whereas we shall find, and indeed can already see from equation 1.10, that the phase difference between the slits gives fringes with the angular scale $\sin\theta \approx \lambda/d$. As long as the slits are well separated, $d \gg d_s$, we have $\sin\theta \ll \sin\theta_s$, and we can safely ignore the slow angular variation due to the finite width of the slits. Secondly, we have assumed that we can simply add the displacements produced by the two separate waves. We shall discuss the conditions under which such an addition is possible at the end of this section; for the moment we concentrate on reducing equation 1.9 to a more transparent form.

We expand the second sine function in equation 1.9, using the appropriate formula from (A 1.2) of appendix 1, to get

$$u = u_1 \sin(\omega t + \eta) + u_2 \cos(\omega t + \eta) \tag{1.11}$$

with

$$u_1 = u_0(1 + \cos\delta) \tag{1.12}$$

and

$$u_2 = u_0 \sin\delta \tag{1.13}$$

Equation 1.11 is typical of the kind of sum that arises in interference problems, and in fact all two-beam interference problems can be reduced to that form. It is convenient to write equation 1.11 as a single sine function; we express it as

$$u = (u_1{}^2 + u_2{}^2)^{1/2}\{\cos\alpha \sin(\omega t + \eta) + \sin\alpha \cos(\omega t + \eta)\} \tag{1.14}$$

where

$$\cos\alpha = u_1/(u_1{}^2 + u_2{}^2)^{1/2} \tag{1.15}$$

with a similar expression for $\sin\alpha$; the removal of the factor $(u_1{}^2 + u_2{}^2)^{1/2}$ ensures that $\sin^2\alpha + \cos^2\alpha = 1$, so that the angle α is properly defined. Finally we see that equation 1.14 is the same as

$$u = (u_1{}^2 + u_2{}^2)^{1/2} \sin(\omega t + \eta + \alpha) \tag{1.16}$$

The advantage of this form is that we have isolated the amplitude $(u_1{}^2 + u_2{}^2)^{1/2}$ and phase α of the resultant displacement.

In the case of the Young's slit experiment, equations 1.12 and 1.14 give for the amplitude u_r of the displacement u

$$u_r = (u_1^2 + u_2{}^2)^{1/2} = u_0\sqrt{2}(1 + \cos\delta)^{1/2} \tag{1.17}$$

What we observe in practice is the intensity I, which is proportional to $u_r{}^2$. For this, equation 1.17 gives

$$I = 2I_0(1 + \cos\delta) \tag{1.18}$$

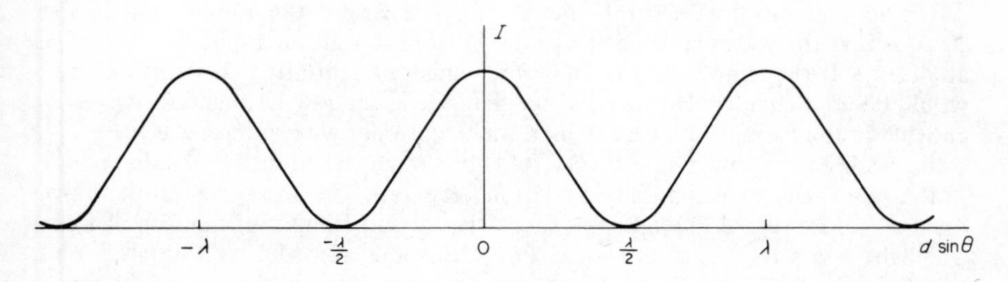

Figure 1.9 Calculated intensity distribution in Young's slits experiment

where I_0 is the intensity associated with a single beam of amplitude u_0. As shown in figure 1.9, I oscillates between $2I_0$ and zero as the path length difference $d \sin \theta$ varies. It can be seen that the graph of the intensity agrees with the fringe pattern shown in figure 1.7. The angular scale of the fringe pattern, for instance the angular distance between successive dark fringes, is $\sin \theta \approx \lambda/d$, as we mentioned earlier.

We may now discuss the conditions under which we may add the amplitudes of two disturbances, as we did in equation 1.9. These conditions are first that the equation of motion for the wave should be linear, and secondly that the waves travelling through the slits S_1 and S_2 should be coherent. We deal with the conditions in turn.

To see what is meant by linearity, we may return to equation 1.1 for the simple harmonic oscillator. We can show that this equation has a property analogous to that assumed in equation 1.9, namely that if $u_1(t)$ is one solution of the equation, and $u_2(t)$ is another, then $u_1(t) + u_2(t)$ is also a solution. For example, $u_1(t)$ might be $\sin \omega t$ and $u_2(t)$ might be $\cos \omega t$; we are then saying that $\sin \omega t + \cos \omega t$ is another possible solution. The proof is simple; we have

$$\begin{aligned} m \frac{d^2}{dt^2}(u_1 + u_2) &= m \frac{d^2 u_1}{dt^2} + m \frac{d^2 u_2}{dt^2} \\ &= -Ku_1 - Ku_2 \qquad (1.19) \\ &= -K(u_1 + u_2) \end{aligned}$$

which is the result we stated. It can be seen that this result holds for any equation of motion that involves only u and its derivatives du/dt, d^2u/dt^2 and so on, and which does not have terms like u^2 or $u(du/dt)$ for example. Such equations are called *linear*, and when we are dealing with a linear equation of motion we may use the *principle of superposition*, as in equation 1.9. The first sign of non-linearity in a medium is that if a wave of a single frequency ω is incident upon it, then other frequencies, typically the first harmonic 2ω, are generated in the medium (see problem 1.8). Harmonic generation does occur with the very intense light from some lasers, but for ordinary light intensities there is no detectable non-linearity. We are therefore on safe ground in assuming that the governing equation for light propagation is linear.

The second condition, that of coherence, is required in order that we may use a fixed phase difference δ between the halves of equation 1.9. Since the phase difference δ is entirely due to the path difference $d \sin \theta$, the implication, at first sight, is that the waves arriving at S_1 and S_2 from the slit S_0 are perfectly in phase at all time. If the lamp G emitted a perfect sine wave, infinite in duration, there would be no difficulty. However, as we shall see in section 4.1, light is always emitted in the form of photons, which for the present we can regard as wave trains lasting for a finite time τ; very roughly τ might be about 10^{-9} s. In an ordinary source, there is no phase relation between successive wave trains, and we must therefore regard the light as a succession of short pulses of random phase. The light is said to have a coherence time τ. In some laser sources, though not all, there is a phase relation between successive wave trains, and the coherence time can be longer. It may perhaps now be asked why any interference effect is observed at all with an ordinary source G. The reason is that the narrow slit S_0, of width d_0 say, imposes an angular spread $\sin \theta_0 \approx \lambda/d_0$ on each wave train, and provided the geometry is such that S_1 and S_2 lie within this angular spread, then parts of each wave train arrive at each slit. This is why S_0 has to be narrow. We say that S_0 ensures that there is *spatial coherence* between S_1 and S_2. While the calculation we have done cannot be regarded as applying to infinite waves, we can now read it as giving the interference between two parts of the same original wave train. If we photograph the interference pattern, with an exposure time of perhaps one second, then we are actually recording the cumulative effect of many wave trains passing through the slits, and the result is the same as if we were indeed dealing with an unending sine wave. The crucial point is that the beams passing through the two slits originate in the same source G.

It is of interest to ask whether it is possible to observe any kind of interference from two slits illuminated by different sources G_1 and G_2. It should be clear from our previous discussion that the answer depends on the nature of the detector D used to record the light. If D has a response time τ_D smaller than the coherence time τ of either source, then D will 'see' the wave trains from G_1 and G_2 as long in duration, and the output of D will be large if the waves of the moment reinforce, small if they cancel. Thus the output of D will flucutate with a characteristic time of τ. If on the other time D has a long response time, like a photographic plate, then no fluctuations will be observed.

The subject of coherence is by no means exhausted by these brief comments. The coherence time τ is a time for loss of 'phase memory' by the light, and one cannot expect that the phase memory is either perfect or non-existent; in general what is required is a 'phase memory function', decaying from 1 to 0 with time. Similarly, two beams need not be fully coherent, or incoherent; the general case is partial coherence. For our purposes, however, it is sufficient to note that interference effects generally require light from a single source.

Worked example 1.2

The following displacements are added together. Find the amplitude of the resulting displacement and its phase relative to $\sin \omega t$ and sketch the two components and the resultant

$$\sin\left(\omega t + \frac{\pi}{3}\right) + \cos\left(\omega t - \frac{\pi}{3}\right)$$

Answer

We expand the sine and cosine, to get

$$u = \sin\left(\omega t + \frac{\pi}{3}\right) + \cos\left(\omega t - \frac{\pi}{3}\right)$$

$$= \tfrac{1}{2} \sin \omega t + \frac{\sqrt{3}}{2} \cos \omega t + \tfrac{1}{2} \cos \omega t + \frac{\sqrt{3}}{2} \sin \omega t$$

since

$$\cos\frac{\pi}{3} = \frac{1}{2} \text{ and } \sin\frac{\pi}{3} = \frac{\sqrt{3}}{2}$$

Collecting terms, we find

$$u = \tfrac{1}{2}(\sqrt{3} + 1)(\sin \omega t + \cos \omega t)$$

To get this into 'amplitude–phase' form, we arrange that the sum of the squares of the coefficients of sin ωt and cos ωt is unity

$$u = \frac{1}{2}\ (\sqrt{3} + 1)\sqrt{2}\left(\frac{1}{\sqrt{2}}\ \sin \omega t + \frac{1}{\sqrt{2}} \cos \omega t\right)$$

$$= \frac{\sqrt{2}}{2}(\sqrt{3} + 1) \sin\left(\omega t + \frac{\pi}{4}\right)$$

sin cos $\pi/4 = \sin \pi/4 = 1/\sqrt{2}$.

The amplitude of the resultant is $\frac{1}{2}\sqrt{2}(\sqrt{3} + 1) = 1.93$ and the phase is $\pi/4$ in advance of sin ωt. As shown in the sketches below, the amplitude of the resultant is nearly 2 because the phases of the two components are not very different.

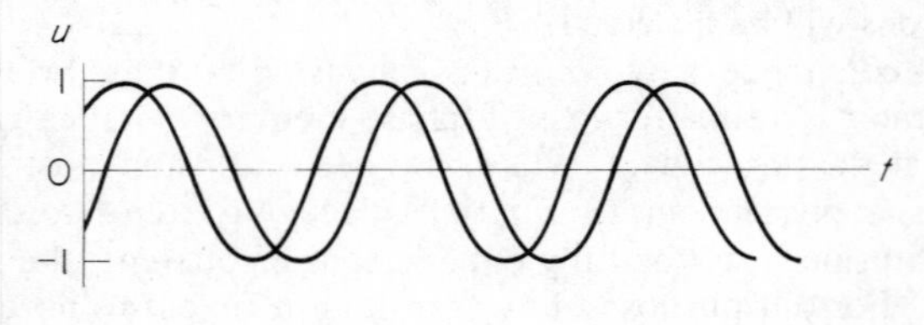

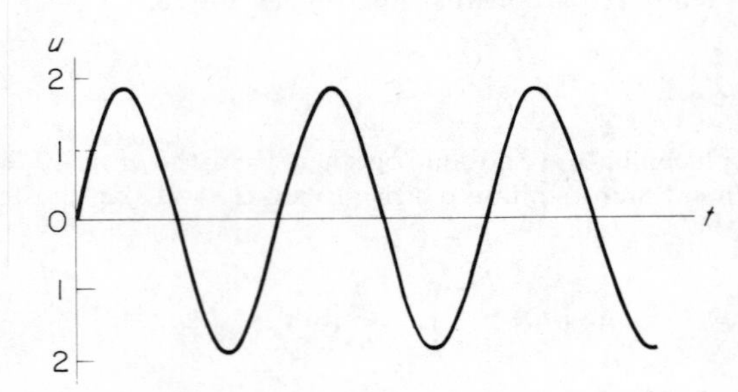

1.3 Beats

We now move on to study the superposition of vibrations of different frequencies; the simplest example is the generation of *beats* when two sound waves of nearly equal frequency interfere. We cannot readily draw an example from optics; as we have seen, even light waves of the same frequency do not generally interfere unless they are from the same source, and interference between light waves of different frequencies from ordinary sources is not possible.

Suppose however we superpose two sound waves of angular frequencies ω_1 and ω_2. For simplicity we take the amplitudes to be equal and the relative phase zero, so that the resulting displacement is

$$u = u_0 \cos \omega_1 t + u_0 \cos \omega_2 t \tag{1.20}$$

Using the formula for the addition of two cosine waves (appendix 1), we can write this as

$$u = 2u_0 \cos \left(\frac{\omega_1 + \omega_2}{2}\right) t \cos \left(\frac{\omega_1 - \omega_2}{2}\right) t \tag{1.21}$$

or

$$u = u_{\text{MOD}}(t) \cos \left(\frac{\omega_1 + \omega_2}{2}\right) t \tag{1.22}$$

with

$$u_{\text{MOD}}(t) = 2u_0 \cos \left(\frac{\omega_1 - \omega_2}{2}\right) t \tag{1.23}$$

The most interesting case is when ω_1 and ω_2 are nearly equal. As sketched in figure 1.10, equation 1.22 then represents an oscillation at the *carrier frequency*

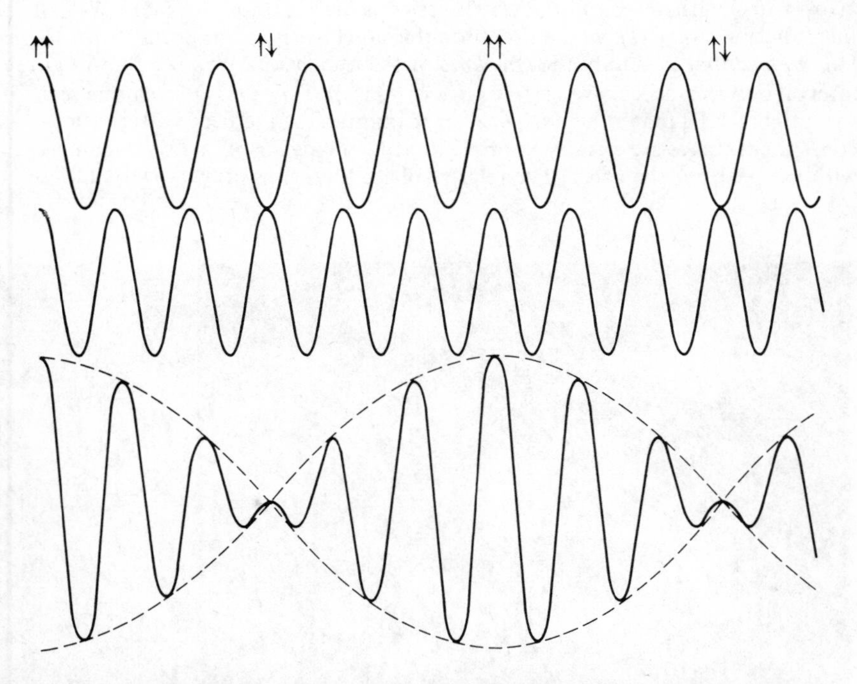

Figure 1.10 Beat envelope formed by the superposition of two waves of nearly equal frequency and equal amplitude. Points where the constituent waves are in phase are marked ↑↑, and points where they are in antiphase are marked ↑↓. The dashed line shows the modulation envelope $u_{\text{MOD}}(t)$. We plot u on the vertical axis and t on the horizontal axis.

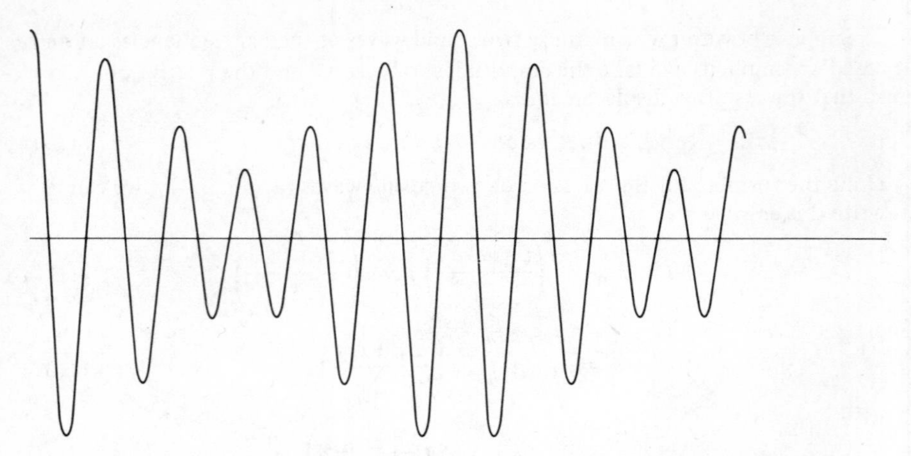

Figure 1.11 Beat envelope formed by the superposition of two waves of nearly equal frequency and unequal amplitude

$\frac{1}{2}(\omega_1 + \omega_2)$ with an amplitude which varies as the function $u_{\text{MOD}}(t)$. We call this function $u_{\text{MOD}}(t)$, which describes the slowly varying amplitude, the *modulation envelope*, or simply the *envelope* of the oscillation. We note that the time interval between successive zeros of $u_{\text{MOD}}(t)$ is $2\pi/(\omega_1 - \omega_2)$. It can be seen from figure 1.10 that at an *antinode*, or maximum, of the beat pattern the constituent waves are exactly in phase, that is, the peaks of one wave coincide with the peaks of the other. The relative phase then slips progressively until at

Figure 1.12 Beat pattern formed by superposing two sheets of Letratone

the next *node*, or zero, of the beat pattern the constituent waves are 180° out of phase with one another.

The effect of superposing waves of unequal amplitude is shown in figure 1.11. The carrier wave and modulation envelope still have the same general form, but the modulation envelope is never zero. The proof is left for problem 1.10.

The formation of beats can easily be demonstrated by feeding two signal generators, tuned to slightly different frequencies, into the same loudspeaker, or by plucking two guitar strings tuned to nearly the same pitch. What is heard is a note of the average frequency $\frac{1}{2}(\omega_1 + \omega_2)$ split into pulses, or beats, of duration $2\pi/(\omega_1 - \omega_2)$. Note that the formation of a beat pattern gives a very accurate measure of the difference between two frequencies; it is easy to measure a difference of one hertz in several hundred hertz in this way. One can also demonstrate beat formation with the aid of sheets of plastic finely ruled with parallel straight lines (for example Letratone). If two sheets of slightly differing ruling are superposed the analogue of a beat pattern is formed, as shown in figure 1.12.

We have deliberately described the beat pattern in the language of radio engineering. In ordinary radio, for example, the signal being transmitted is at audio frequency, the frequency of ordinary sound, say up to 10^4 Hz. The a.f. signal is transported as a modulation of the radio frequency, or r.f., carrier wave, of frequency typically 10^6 Hz. The beat pattern, figure 1.10, is the simplest example of *amplitude modulation* of a carrier wave.

1.4 General Waveforms and Fourier Analysis

In setting up the beat pattern, we combined two single-frequency vibrations. It is instructive to represent the operation as in figure 1.13, where we show the waveform both in the *time domain u* versus t, and in the *frequency domain*, $|\tilde{u}|$ versus ω. The meaning of the latter graph, which is new, is that we show the amplitude of the component vibrations, in this case simply equal amplitudes at ω_1 and ω_2. Note that as we are showing it, the frequency-domain plot does not contain any information about the relative phase of the two waves, so that it is less complete than the time-domain plot. In particular, we always plot the amplitude as positive, and it is for this reason that we write the amplitude as $|\tilde{u}|$. We may say that in investigating the beat pattern we used synthesis, in that we started from a knowledge of the frequency-domain form $|\tilde{u}|$ versus ω, and worked out the

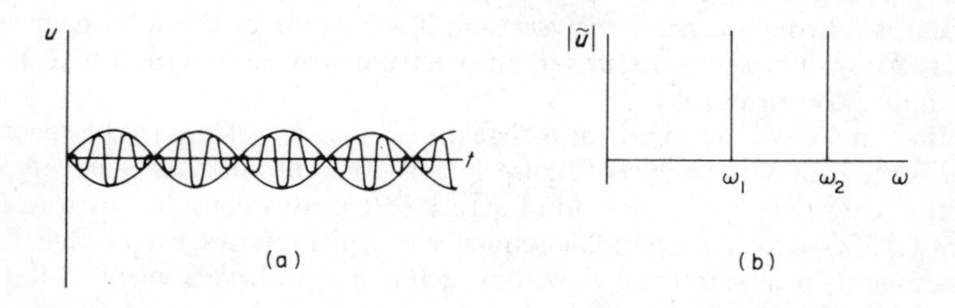

Figure 1.13 Beat pattern in time and frequency domains

behaviour in the time domain, u versus t. We now turn to the opposite question, that of analysis: suppose we are given a waveform in time, $u(t)$, what is the corresponding frequency spectrum, $|\tilde{u}(\omega)|$? This question is of obvious practical

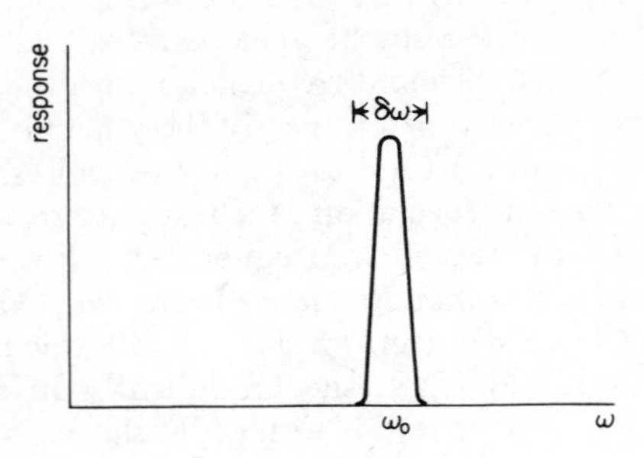

Figure 1.14 Narrow band response of a radio set

importance in communications engineering, since the answer to it determines the frequency bandwidth which is required to pass the given signal without distortion; the appropriate bandwidth is the width in ω over which $|\tilde{u}(\omega)|$ is non-zero. More generally, we shall see that an understanding of the relationship between $u(t)$ and $|\tilde{u}(\omega)|$ illuminates many aspects of our subject.

One can tackle the problem of the connection between a wave form $u(t)$ and the corresponding $|\tilde{u}(\omega)|$ in a purely mathematical way. The approach is known as *Fourier analysis* and $|\tilde{u}(\omega)|$ is called the *Fourier transform* of $u(t)$. We shall find it more useful, however, to adopt an experimental approach. To begin with, consider an ordinary radio set, which is a narrow-band detector; that is, when the radio is tuned to a station, it is detecting the amplitude of the r.f. signal falling on the aerial at the frequency of that particular station. More exactly, the set detects the strength of signal in a narrow band $\delta\omega$ about the frequency ω_0 to which it is tuned, as shown in figure 1.14. As the tuning is altered, the central frequency ω_0 is moved up or down, and by moving ω_0 through a range of frequencies we can plot out the amplitude of the signal versus frequency in that range. Thus the radio set can be used to measure what we have called $|\tilde{u}(\omega)|$. It is possible to arrange for the frequency to be swept automatically; the resulting instrument is called a *spectrum analyser,* and it gives a direct oscilloscope reading of $|\tilde{u}(\omega)|$. We shall base our discussion on spectrum analyser traces of $|\tilde{u}(\omega)|$ for various input waveforms $u(t)$.

The first, and obvious, comment is that if the input waveform to the spectrum analyser is the beat pattern $u(t)$ of figure 1.13(a), the output $|\tilde{u}(\omega)|$ indeed consists of the two single spikes shown in figure 1.13(b). Now consider the waveform of figure 1.15(a), which is a periodic sequence of square pulses, each of length t_0 and separated by a time interval T. Within each pulse, we have a segment of a sine wave, which if it were infinite would correspond to a frequency ω_0. The frequency spectrum is shown in figure 1.15(b); our waveform is made up of a superposition

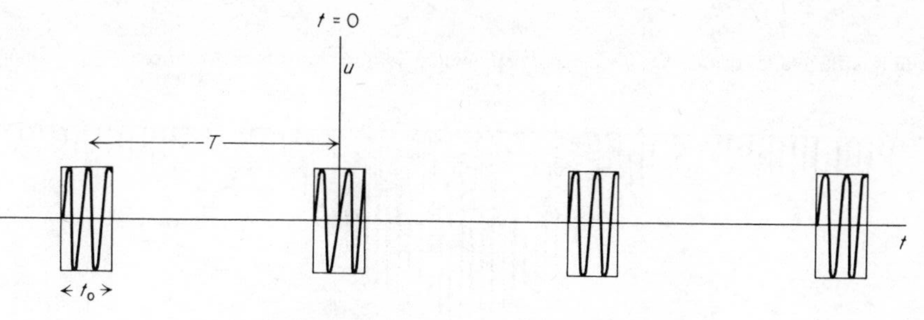

Figure 1.15(a) Sequence of square pulses of duration t_0

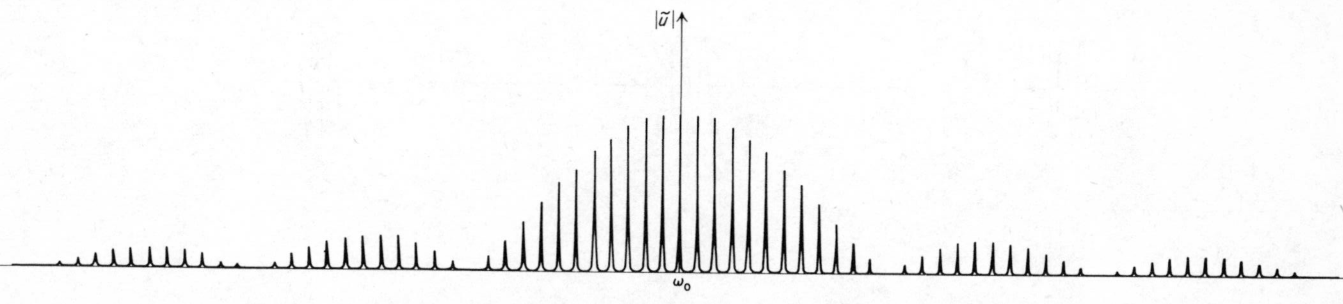

Figure 1.15(b) Frequency spectrum of pulse train in figure 1.15(a)

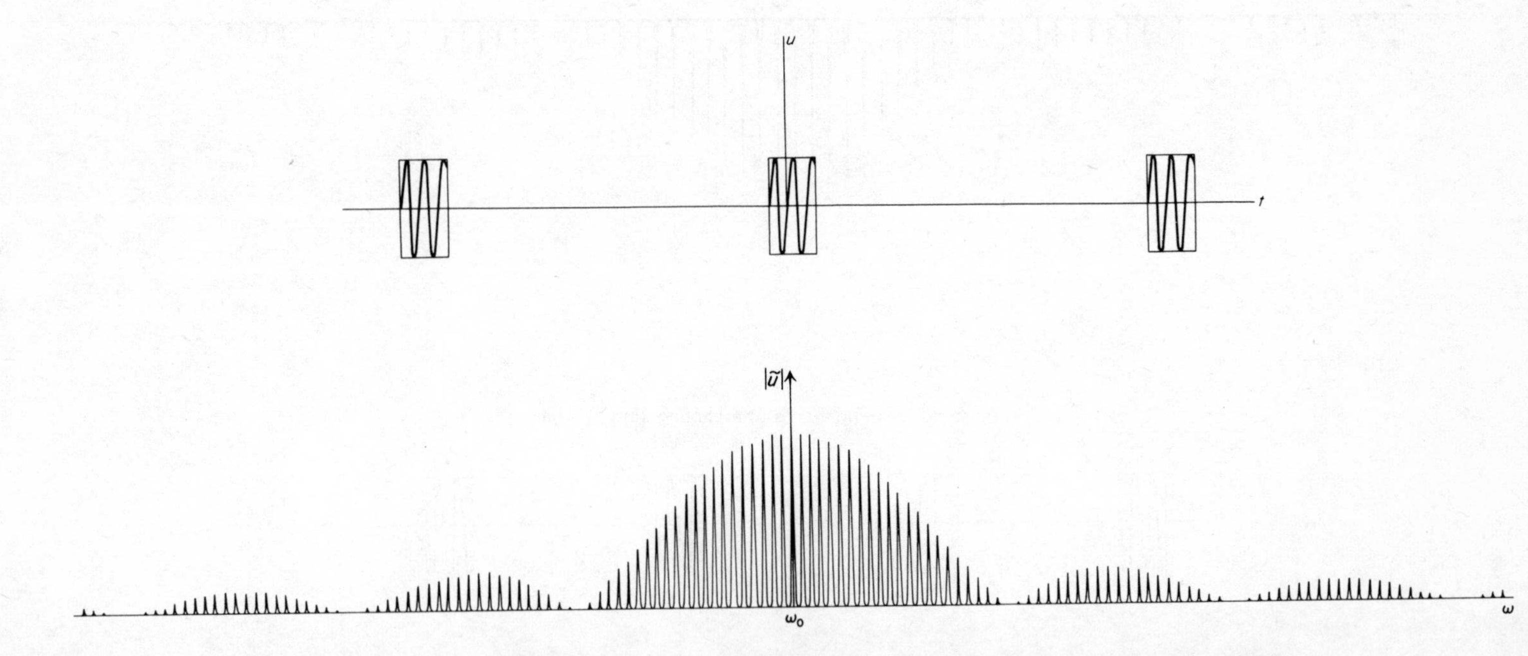

Figure 1.16 Frequency spectrum of pulse train shown in figure 1.15(a) with repeat time T increased

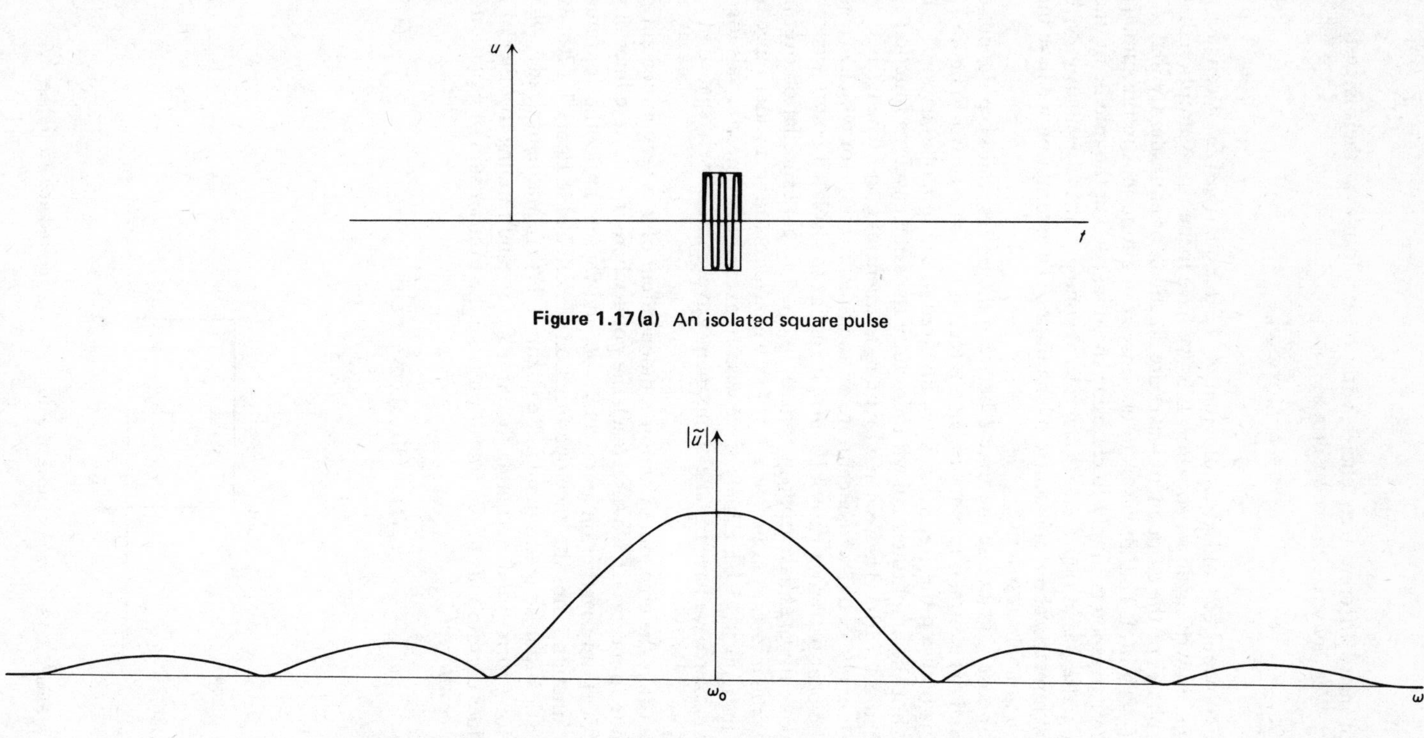

Figure 1.17(a) An isolated square pulse

Figure 1.17(b) Frequency spectrum of the isolated square pulse of figure 1.17(a)

of waves of many different frequencies, with the amplitudes at their largest around ω_0. We may write the waveform as

$$u(t) = \sum_n u_n \cos \omega_n t \tag{1.24}$$

which is an obvious generalisation of equation 1.20. Equation 1.24 is called a *Fourier sum.* In writing down equation 1.24 we have made the assumption that the initial phases of the component waves are all equal, or equivalently that $u(t)$ is even, $u(-t) = u(t)$. This allows us to use only cosine functions in the sum. The assumption corresponds simply to choosing the time $t = 0$ in the middle of one of the square pulses, as in figure 1.15(a), and to assuming that the oscillation within a pulse is symmetric about the centre of the pulse. The assumption we have made is therefore not very restrictive.

Now suppose we increase the time T between the pulses, and keep the pulse length t_0 and the central frequency ω_0 unaltered. The result is shown in figure 1.16. The spacing between the frequencies ω_n involved in the sum of equation 1.24 decreases, but the amplitudes fall within exactly the same envelope function as those in figure 1.15(b). The reason why more frequencies occur when T is increased is easily seen if we concentrate on the time intervals for which $u(t)$ is zero. In the beat pattern, figure 1.10, $u(t)$ is zero at the isolated points where the two component waves have opposite phase. In order to get cancellation over more extended time intervals, as in figure 1.15(a), we have to add in a wider range of waves, as in figure 1.15(b). Finally if we wish to extend the time intervals for which $u(t)$ is zero, we have to take component waves more closely spaced in frequency.

The fact that the envelope function is independent of T is very important. If we continue to increase T, we soon reach the point where the discrete lines of figure 1.15(b) run together on an oscilloscope, and we see simply the continuous envelope function. Thus the envelope function as sketched in figure 1.17(b) is characteristic of an isolated square pulse (figure 1.17(a)). Mathematically, what we have done corresponds to taking the limit $T \to \infty$, and running together of the discrete lines corresponds to the transition from the Fourier sum of equation 1.24 to a *Fourier integral*

$$u(t) = \int \tilde{u}(\omega) \cos \omega t \, d\omega \tag{1.25}$$

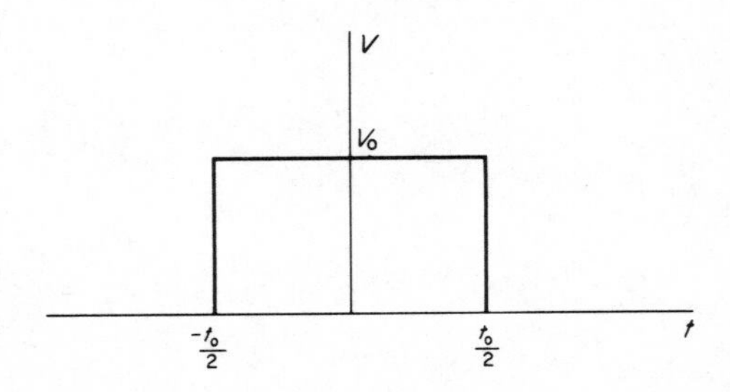

Figure 1.18 A square pulse of amplitude V_0, centred at zero frequency

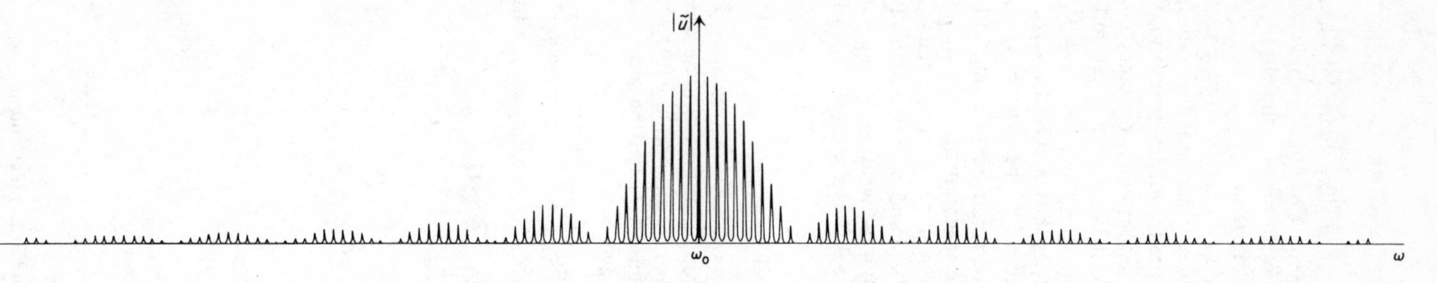

Figure 1.19 Frequency spectrum of a train of pulses of duration $2t_0$

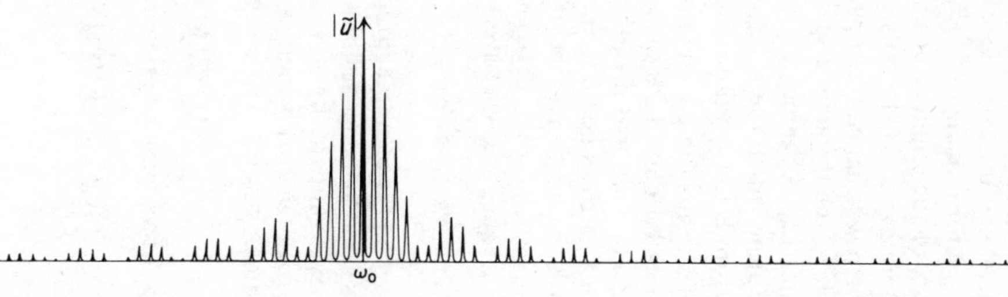

Figure 1.20 Frequency spectrum of a train of pulses of duration $4t_0$

Thus a Fourier sum corresponds to a repeated waveform, like figure 1.15(a) or 1.10 and a Fourier integral corresponds to an isolated waveform, like figure 1.17(a). We shall mainly concentrate on isolated waveforms $u(t)$ and the characteristic envelope functions $|\tilde{u}(\omega)|$ that go with them.

The square waveform of figure 1.17(a) may be described by two parameters: the frequency ω_0 of the segment of a sine wave which is contained within the pulse, and the duration t_0 of the pulse. How does the frequency spectrum, figure 1.17(b), alter when we change these parameters? First, as might be expected, if we alter ω_0, say to ω_0', the frequency spectrum retains its shape, but is centred on the new frequency ω_0'. An important special case is $\omega_0' = 0$; we then have simply a square pulse during which the voltage is constant, as shown in figure 1.18. The frequency spectrum is the envelope function of figure 1.17(b), centred at $\omega_0 = 0$.

The effect of altering the duration t_0 of the pulse is more interesting and more important. Figures 1.19 and 1.20 show appropriate spectrum analyser traces, from which it can be seen that if t_0 increases, the envelope function retains its general shape, but narrows down to include a smaller band of frequencies. This also happens with the beat envelope of figure 1.10; if we write δt for the duration of each beat, namely $2\pi/(\omega_1 - \omega_2)$, and $\delta\omega$ for the difference $\omega_1 - \omega_2$ of the two component frequencies, we have

$$\delta\omega\delta t = 2\pi \qquad \text{(beat envelope)} \tag{1.26}$$

Thus, if δt is doubled for example, $\delta\omega$ is halved. We can see from equation 1.24 or 1.25 that this same 'halving-doubling' relationship holds for a general waveform. In fact, suppose we compare the waveform generated by a function $\tilde{u}(\omega)$ with that generated by $\tilde{u}_2(\omega) = \tilde{u}(2\omega)$, which as shown in figure 1.21 has the same shape but half the frequency range. The corresponding waveform is

$$u_2(t) = \int \tilde{u}(2\omega) \cos \omega t \, \mathrm{d}\omega \tag{1.27}$$

and by changing the variable of integration from ω to 2ω, we may write this as

$$u_2(t) = \tfrac{1}{2} \int \tilde{u}(\omega_1) \cos \tfrac{1}{2}\omega_1 t \, \mathrm{d}\omega_1 = \tfrac{1}{2}u(\tfrac{1}{2}t) \tag{1.28}$$

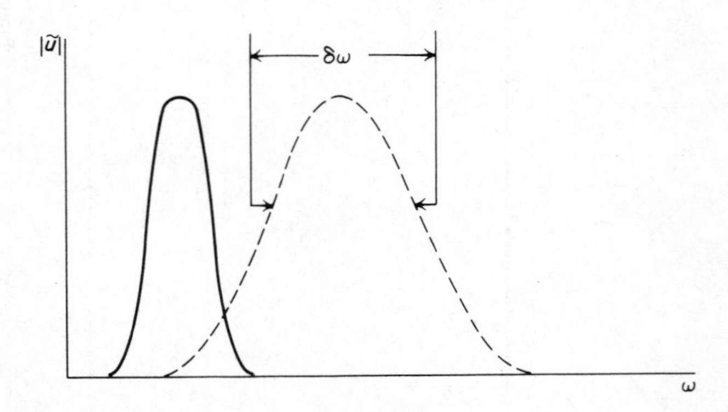

Figure 1.21 Two functions $\tilde{u}(\omega)$ (dotted line) and $\tilde{u}(2\omega)$ (continuous line)

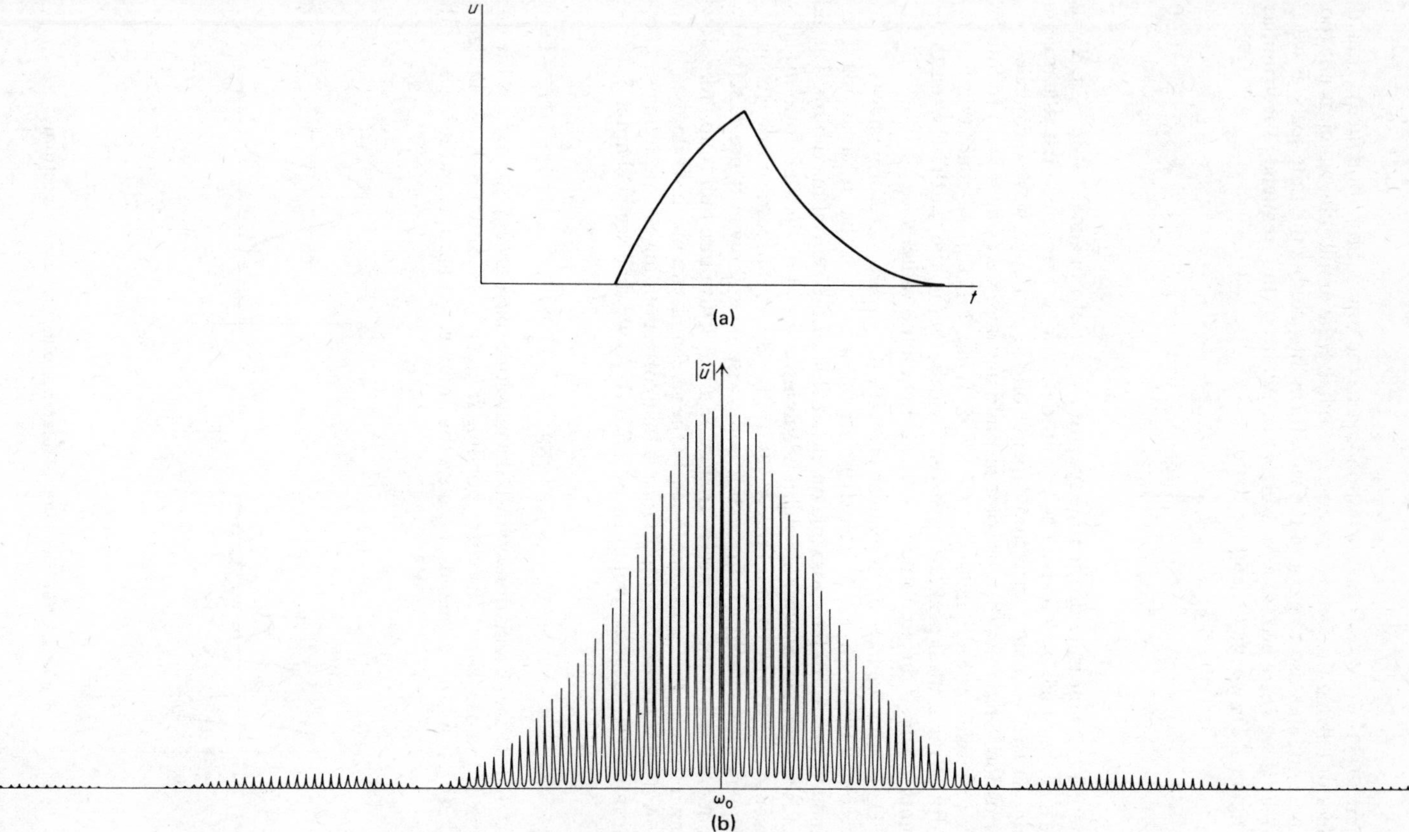

Figure 1.22(a) Triangular pulse and (b) corresponding frequency spectrum

where $u(t)$ is the waveform corresponding to the original function $\tilde{u}(\omega)$. Thus the effect of halving the frequency range is to halve the amplitude, which is not too important, and also to expand the waveform into double the time span, as we stated. If we write $\delta\omega$ for a measure of the width of the frequency spectrum, as in figure 1.21, we may therefore write

$$\left.\begin{array}{l}\delta\omega\delta t \approx 2\pi \\ \text{or} \\ \delta f \delta t \approx 1\end{array}\right\} \text{(general waveform)} \qquad (1.29)$$

where δt is a measure of the time span of the waveform and $\delta f = \delta\omega/2\pi$. We use the sign $\approx$, which may be read as 'is of order', for two reasons. First, the precise shape of the frequency spectrum depends on the particular waveform under discussion. Secondly, there is some latitude in where we calliper $\tilde{u}(\omega)$ to find $\delta\omega$. For example, in the square-wave spectrum, figure 1.17(b), we could take $\delta\omega$ as the width of the central peak, or as the half width, that is, the width between the points at which the amplitude is half the maximum value.

Up to now we have dealt exclusively with the square waveform and its frequency spectrum, the most striking aspect of which is that it has strong sidebands, that is subsidiary maxima on either side of the central maximum. The sidebands are connected with the sharp beginning and end of the waveform $u(t)$, because in order to build up the waveform rapidly it is necessary to have high-frequency components in $\tilde{u}(\omega)$. For comparison, we show in figure 1.22(b) the frequency spectrum of a nearly triangular pulse. This pulse builds up more slowly than a square pulse, and consequently the sidebands are much weaker.

A third waveform of importance is that corresponding to a decaying simple harmonic oscillator, shown in figure 1.23(a). As we shall see in chapter 5, the precise form is

$$u(t) = \exp(-t/2\tau) \cos \omega_1 t \qquad (1.30)$$

which describes oscillations within the envelope $\exp(-t/2\tau)$ so that τ is the decay time of the oscillator. The corresponding frequency spectrum is shown in figure 1.23(b). It has no sidebands, because the decay is not abrupt, and it has a width

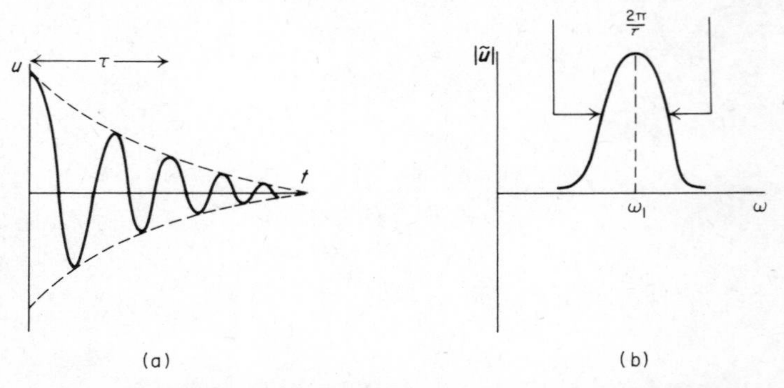

Figure 1.23(a) Decaying oscillator and (b) frequency spectrum

$2\pi/\tau$ as marked. Note that unlike the square and triangular pulses, the waveform of figure 1.23(a) is not an even function; that is, one that satisfies $u(-t) = u(t)$. Consequently, it cannot be represented simply by a sum over cosine functions, like equation 1.24; terms in $\sin \omega_n t$ are required as well. As we remarked at the beginning of this section, this means that $|\tilde{u}(\omega)|$ shows the total amplitude at frequency ω, calculated from the separate sine and cosine amplitudes as in the progression from equation 1.11 to 1.16.

1.5 Illustrations of Uncertainty Relation

The most important result of the previous section is equation 1.29 relating the duration δt of a waveform $u(t)$ to the width $\delta\omega$ of the frequency spectrum $\tilde{u}(\omega)$. For historical reasons, which will become clearer in chapter 4, we shall refer to equation 1.29 as the uncertainty relation. The relation is widely used to find order-of-magnitude estimates of frequency widths. Perhaps the most common application is to the properties of an unstable particle, or of a slowly decaying excitation of a solid. Equation 1.29 is used to express the decay time, or half life, of the particle as a linewidth in frequency. A second application is to diffraction patterns. It is possible to build up waveforms in space rather than in time, and we shall see in chapter 3 that the spatial form of equation 1.29 determines the characteristics of all diffraction patterns. However, although the central importance of equation 1.29 will become clear later in the book, it is worth while giving two illustrations of its use straightaway.

First, consider the emission of radiation by the atoms in a gas—for example, let us say neon in a discharge tube. The radiation emitted is the familiar red of a neon tube, centred on a wavelength of 633 nm. The radiation emitted by an isolated atom is a sharp spectral line. In the gas, however, the line is broadened,

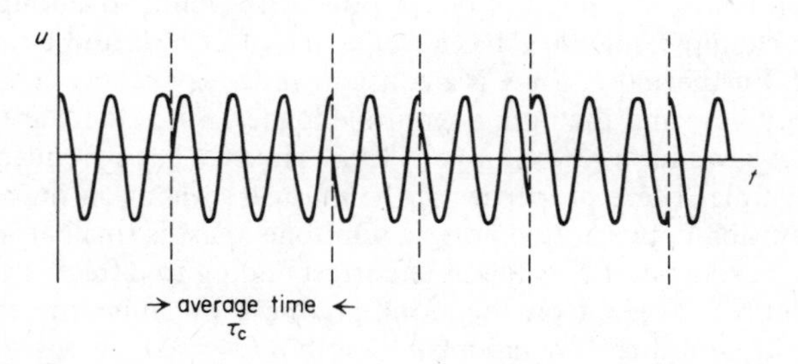

Figure 1.24 Collision broadening

first by the Doppler effect, because some of the atoms are moving towards and some away from the observer, and secondly by the process known as collision broadening, which is what concerns us now. The atoms in the gas undergo collisions with each other, and during each collision the radiation emitted by an atom changes phase abruptly, as illustrated in figure 1.24. We have already

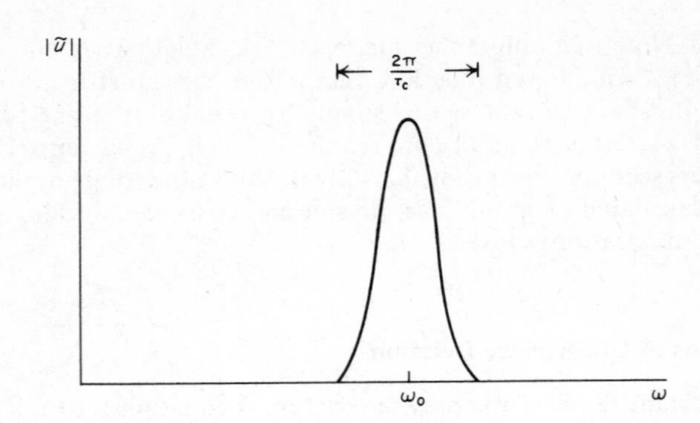

Figure 1.25 Collision broadened line

mentioned, towards the end of section 1.2, that the isolated atom emits light as a stream of photons, each of which may be regarded as a light train lasting for a time τ. The effect of collisions is that in a gas the light is emitted in pulses of average duration τ_c, the mean time between collisions, and τ_c can be much shorter than τ. Consequently the spectral line is broadened to a linewidth $2\pi/\tau_c$ as in figure 1.25. The Doppler broadening is determined by the mean velocity of the atoms, that is, by the temperature, and it is independent of the number of atoms present. That is, the Doppler broadening is independent of the pressure of the gas. The collision breadth, on the other hand, increases with pressure, since the mean time τ_c between collisions decreases as the pressure increases. Consequently Doppler broadening is the dominant mechanism at low pressure, and collision broadening at high pressure. Some experimental results are shown in figure 1.26.

For our second example, we may turn to one of the difficulties faced by the player of a low-pitched musical instrument. The musical scale has the basic property that an increase in pitch of one octave corresponds to a doubling of frequency. For example, middle C has a frequency of 261 Hz and C an octave higher, 522 Hz. Furthermore, there is a constant ratio, say r, between the frequency of any note and the note a semitone higher. Since there are 12 semitones in an octave, we have $r^{12} = 2$, or $r = 1.06$. Thus C$^\#$ has a frequency of 277 Hz, for example. These properties of the musical scale mean that the frequency separation between two notes a semitone apart is small at low frequencies. For example, if N is the note corresponding to a frequency of 50 Hz, N$^\#$ is 53 Hz, and N$^\flat$, 47 Hz. Now the 'note' produced by an instrument is of finite duration δt, and therefore has a bandwidth $\delta f \approx 1/\delta t$. As shown in figure 1.27, for the note to have a well-defined pitch, we may expect that δf should be smaller than the frequency difference between N$^\#$ and N

$$\delta f < 0.06 f \tag{1.31}$$

and therefore

$$\delta \tau > 1/0.06 f \tag{1.32}$$

For $f = 50$ Hz, this gives $\delta\tau = \frac{1}{3}$ s as the minimum time for which a note may have to be sustained in order to give a more or less definite pitch. Thus the uncertainty relation gives a fundamental physical constraint on the speed with which passages

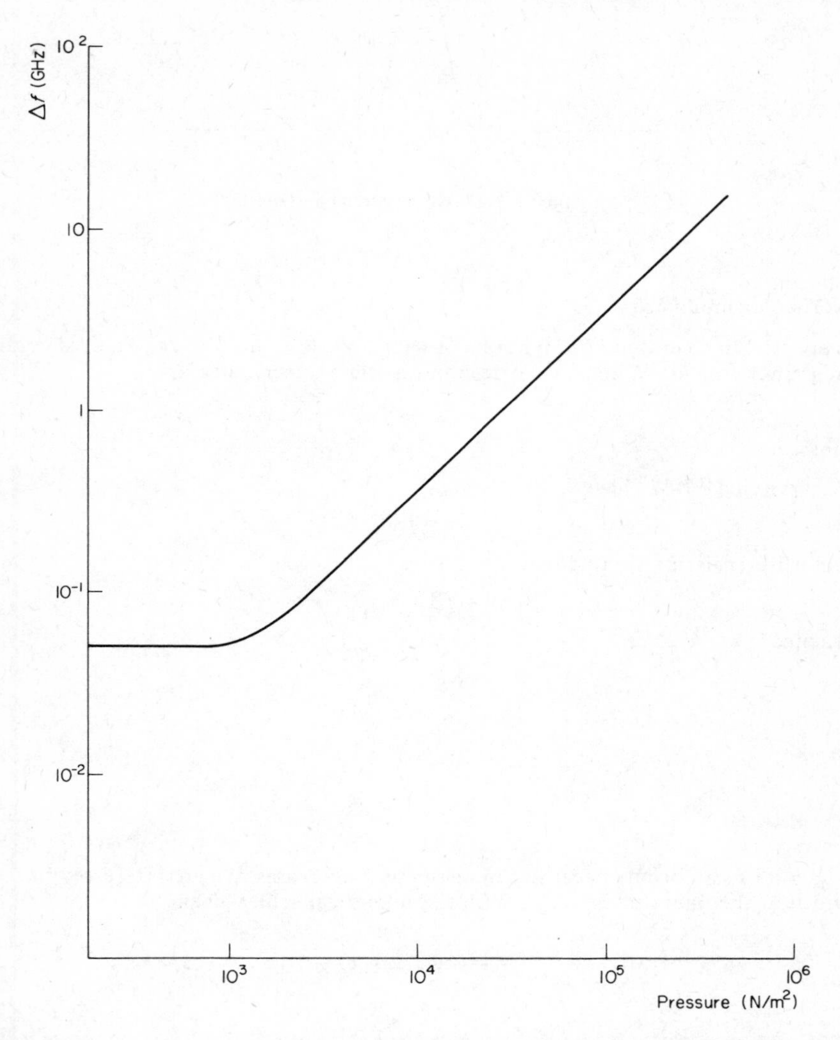

Figure 1.26 Linewidth as a function of pressure in CO_2 at room temperature. The linewidth is constant at low pressure—this is the region of Doppler broadening—and then increases with pressure as collision broadening takes over. (Courtesy of Professor A. F. Gibson)

of notes can be played at low pitch. At a high frequency, on the other hand, the uncertainty relation does not impose any constraint. For example, at $f = 1$ kHz equation 1.32 gives $\delta\tau > 0.017$ s, which is obviously of no consequence.

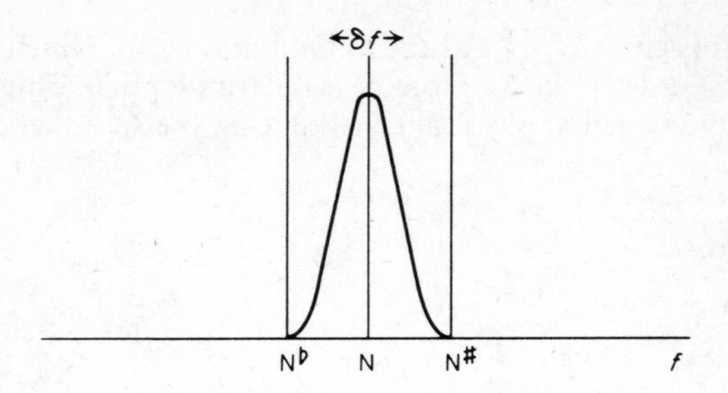

Figure 1.27 Note with a bandwidth

Worked example 1.3

A spectral line centred at 550 nm has a width of 10^{-3} nm. Find the corresponding frequency spread δf, and the corresponding time uncertainty δt.

Answer

First we find δf. We have

$$c = f\lambda$$

and differentiating we find

$$0 = f\delta\lambda + \lambda\delta f$$

whence

$$\frac{\delta\lambda}{\lambda} = -\frac{\delta f}{f}$$

or

$$\delta f = -\frac{f}{\lambda}\,\delta\lambda$$

The minus sign occurs because f increases as λ decreases. We are interested in the width of the line, that is, $|\delta\lambda|$. With the numerical values given

$$\frac{\delta\lambda}{\lambda} = \frac{10^{-3}}{5.5 \times 10^{2}} = 1.82 \times 10^{-6}$$

$$f = \frac{c}{\lambda} = \frac{3 \times 10^{8}}{5.5 \times 10^{-7}}$$

$$= 5.45 \times 10^{14}\ \text{Hz}$$

Hence

$$|\delta f| = f\frac{\delta\lambda}{\lambda} = 9.9 \times 10^{8}\ \text{Hz}$$

The corresponding time uncertainty is

$$\delta t = \frac{1}{|\delta f|}$$
$$= 1.01 \times 10^{-9} \text{ s}$$

Problems

1. The total voltage drop around the ringing circuit of figure 1.4 is zero. Show that this implies that

$$L \frac{d^2 Q}{dt^2} + \frac{1}{C} Q = 0$$

and hence that the circuit obeys equation 1.1.

2. (a) Show, by differentiating twice, that $u = A \sin(\omega t + \eta)$ does satisfy the harmonic oscillator equation

$$m \frac{d^2 u}{dt^2} = -Ku$$

where $\omega^2 = K/m$, as stated in the text.

(b) Give three examples of simple harmonic oscillation. Is simple harmonic oscillation a common occurrence in natural systems?

3. Prove that in simple harmonic motion if the initial displacement is u_0 and the initial velocity is v_0, the amplitude is $(u_0{}^2 + v_0{}^2/\omega^2)^{1/2}$ and the initial phase in advance of $\cos \omega t$ is $-\tan^{-1} v_0/\omega u_0$.

4. A body is suspended from a wire attached to a rigid support. The body is set into torsional oscillations about the suspension as axis. Assume a restoring couple of moment C exists in the wire and that the moment of inertia of the body about the axis of twist is I. Show that the oscillations are simple harmonic and derive an expression for the angular frequency ω. What are the SI units for C, I and ω?

5. An inductor of 100 μH and a capacitor of 10 pF are wired in series. What is the frequency of oscillation?

6. The following sets of displacements are added together. In each case find the amplitude of the resulting displacement and its phase relative to $\sin \omega t$ and sketch the two components and the resultant.

(a) $2 \cos \omega t + 3 \sin \omega t$
(b) $\cos \omega t + 2 \sin \omega t$
(c) $\cos(\omega t + \pi/3) + \cos(\omega t - \pi/3)$
(d) $\sin \omega t + 3 \sin(\omega t + \pi/4)$

7. Consider the non-linear equation

$$a \frac{d^2 u}{dt^2} = bu^2$$

Show that if u_1 and u_2 are solutions, then $u_1 + u_2$ is not a solution.

8. Weak non-linearity. Consider the equation of motion

$$m \frac{d^2 u}{dt^2} = -Ku + \epsilon u^2$$

where ϵ is small. Write the solution as

$$u = u_0 \sin \omega t + \epsilon u_1(t)$$

in which the first term is a solution of the linear equation with $\epsilon = 0$ and $\epsilon u_1(t)$ is a small correction. Show that, neglecting terms in ϵ^2, u_1 satisfies the equation

$$m \frac{d^2 u_1}{dt^2} + Ku_1 = u_0{}^2 \sin^2 \omega t$$

$$= \tfrac{1}{2} u_0{}^2 (1 - \cos 2\omega t)$$

Show (by direct substitution or otherwise) that a solution for u_1 is

$$u_1 = \tfrac{1}{2} u_0{}^2 \left(\frac{1}{K} + \frac{1}{4m\omega^2 - K} \cos 2\omega t \right)$$

so that u_1 contains a d.c. component and a component at the first harmonic frequency 2ω. This is an example of harmonic generation by a non-linear medium.

9. Suppose we have an initial relative phase when we set up a beat pattern, so that equation 1.20 is replaced by

$$u = u_0 \sin \omega_1 t + u_0 \sin (\omega_2 t + \eta)$$

Show that the beat pattern retains the form of equation 1.21, except that the initial phase angles of the carrier and modulation envelope are altered.

10. Suppose we set up a beat pattern between waves of unequal amplitude

$$u = u_1 \sin \omega_1 t + u_2 \sin \omega_2 t$$

Show that the displacement may be written as

$$u = (u_1 + u_2) \sin \frac{\omega_1 + \omega_2}{2} t \cos \frac{\omega_1 - \omega_2}{2} t$$

$$+ (u_1 - u_2) \cos \frac{\omega_1 + \omega_2}{2} t \sin \frac{\omega_1 - \omega_2}{2} t$$

and that this corresponds to figure 1.11.

11. The 643 nm cadmium line has a width of about 7×10^{-4} nm. If this is interpreted as being due to the emission of a succession of wavetrains of limited duration δt, as in collision broadening, calculate δt and the number of cycles in each wavetrain. (Note: to calculate δt you need the frequency bandwidth $\delta\omega$. You are given the wavelength width $\delta\lambda$ and going from $\delta\lambda$ to $\delta\omega$ requires care.)

2

General Properties of Waves

2.1 Wave Number and Wave Velocity

In chapter 1 we discussed waveforms as a function of time; we now turn to the more general question of the propagation of a wave through space, so that we must deal with a displacement which is a function of both space and time. We shall restrict our attention mainly to a wave which is travelling in a fixed direction, say along the x axis, so the displacement is $u(x, t)$. As we mentioned in chapter 1, the nature of the displacement u depends on the character of the wave in question: for a wave on a stretched string (figure 1.1) u is the transverse displacement, for a light wave u may be taken as the local value of the electric field, and so on. We shall discuss specific examples of wave motion later in this chapter; in this section and the next we deal with the basic notation.

To begin with, we can simplify matters by letting $u(x, t)$ be a sine wave

$$u(x, t) = A \sin(kx - \omega t - \delta) \tag{2.1}$$

There are two reasons why this simplification is desirable. First, in general the propagation velocity of a sine wave is a function of the angular frequency ω, so it is best to discuss the single sine wave first. Secondly, just as equation 1.24 expresses a general displacement $u(t)$ as a sum of sinusoidal oscillations, so we can build up a general displacement $u(x, t)$ by superposition of sinusoidal waves. The properties of the sine wave are therefore fundamental.

In equation 2.1, the parameters A, ω and δ, respectively amplitude, angular frequency and phase angle, are already familiar. To find the significance of the

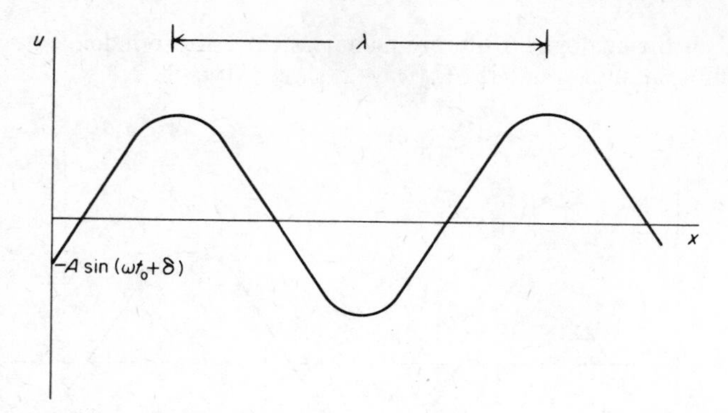

Figure 2.1 The wave of equation 2.1 at time t_0, $u(x, t) = A \sin(kx - \omega t_0 - \delta)$.

parameter k, we may sketch the wave at some instant t_0, as in figure 2.1. It can be seen that when x increases by the wavelength λ, the argument of the sine function increases by 2π. This means $k\lambda = 2\pi$, that is,

$$k = 2\pi/\lambda \tag{2.2}$$

This is analogous to equation 1.6 relating ω to the periodic time T and indeed it can be seen from equations 2.1 and 2.2 that k stands in relation to the spatial coordinate x just as ω stands in relation to the time t. k is called the *angular wave number.* Note in particular that small k corresponds to long wavelength λ.

The correspondence between k and x on the one hand, and ω and t on the other, means that we could parallel section 1.4 with a discussion of the synthesis of a general waveform $u(x)$ in space from sinusoidal waves. The main point of interest would be the relationship between $u(x)$ and the corresponding amplitude function $|\tilde{u}(k)|$, and figures like 1.22 and 1.23 could simply be relabelled as pairs $u(x)$ and $|\tilde{u}(k)|$, rather than $u(t)$ and $|\tilde{u}(\omega)|$. Finally, equation 1.29 may be complemented by its *spatial equivalent.*

$$\delta k \delta x \approx 2\pi \tag{2.3}$$

which relates the widths δx and δk of the functions $u(x)$ and $|\tilde{u}(k)|$. We shall see in chapter 3 that diffraction of waves is governed by equation 2.3.

Waves are sometimes characterised by the ordinary wave number K,

$$K = 1/\lambda \tag{2.4}$$

Table 2.1

Time	Angular frequency ω	Periodic time $T = 2\pi/\omega$	Frequency $f = \omega/2\pi$
Space	Angular wave no. k	Wavelength $\lambda = 2\pi/k$	Wave no. $K = k/2\pi$

Obviously K is the analogue of the frequency f. The correspondence between the temporal and spatial parameters of a wave is shown in table 2.1.

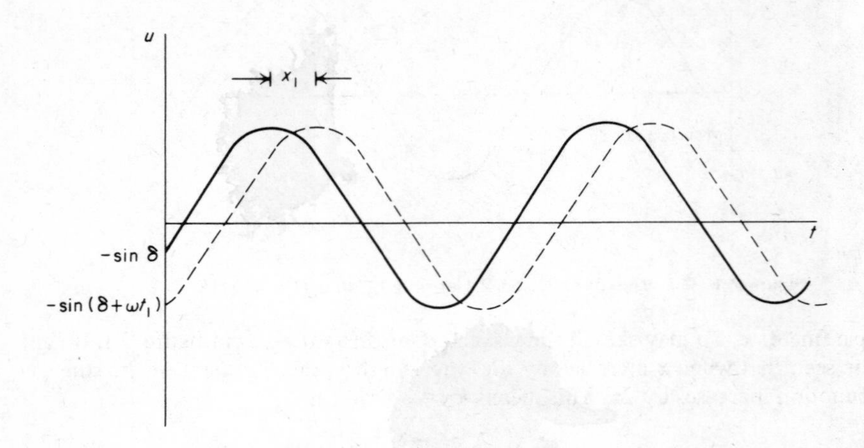

Figure 2.2 Wave of equation 2.1 at times $t = 0$ and $t = t_1$

We can readily find the velocity of propagation of the wave described by equation 2.1. Suppose we look at u as a function of x at two different times $t = 0$ and $t = t_1$. As shown in figure 2.2, the waveform at the later time is the same as that at the earlier time, but shifted to the right a distance x_1, where $kx_1 = \omega t_1$. Thus the wave is travelling to the right with a velocity $v = x_1/t_1$, or

$$v = \omega/k \tag{2.5}$$

We can use table 2.1 to convert this to the form

$$v = f\lambda \tag{2.6}$$

which is perhaps better known. We can see by the same argument that the wave

$$u(x, t) = A \sin(kx + \omega t - \delta) \tag{2.7}$$

travels to the left with the velocity v of equation 2.5.

2.2 Standing Waves and Laser Cavities

The waves of equations 2.1 and 2.7 travel to the right and left respectively; they are called *running waves.* At the opposite extreme is the *standing wave*, which for example may be set up on a string stretched between two fixed points. As shown in figure 2.3, a standing wave does not move, but vibrates with a displacement which is periodic in time. Furthermore, the wavelength of the vibration cannot have any value, but must bear a simple relation to the distance L between the fixed points. For example, in figure 2.3, $\lambda = 2L/3$.

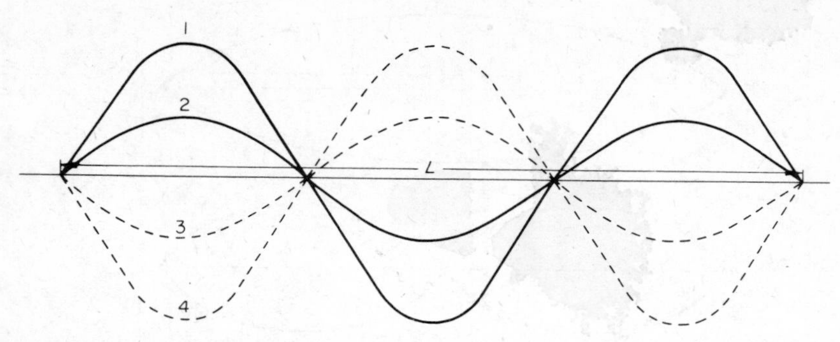

Figure 2.3 Standing wave on a stretched string. The string passes successively through the shapes marked 1, 2, 3, 4 and back again

We may find an expression for the displacement $u(x, t)$ in a standing wave by regarding it as a superposition of two waves of equal amplitude travelling in opposite directions. This gives

$$u(x, t) = A \sin(kx - \omega t) + A \sin(kx + \omega t)$$
$$= 2A \sin kx \cos \omega t \tag{2.8}$$

To see that this displacement does correspond to something like figure 2.3, we may write it as

$$u(x, t) = A(t) \sin kx \tag{2.9}$$

with the time varying amplitude $A(t)$ given by

$$A(t) = 2A \cos \omega t \tag{2.10}$$

Just as sketched in figure 2.3, the shape of the wave in space is given by the factor $\sin kx$, and the amplitude $A(t)$ oscillates with frequency ω. At the points x_n given by

$$kx_n = n\pi \tag{2.11}$$

$\sin kx$ is zero, and therefore the displacement is always zero; these points are called the *nodes* of the standing wave. Conversely, at the points

$$kx_a = (n + \tfrac{1}{2})\pi \tag{2.12}$$

$\sin kx$ is ± 1, and the displacement oscillates between $+2A$ and $-2A$; these points are called the *antinodes*.

As stated previously, in a standing wave the wavelength λ, and therefore the angular wave number k, cannot take *any* value. For example, in a standing wave on a stretched string of length L, the displacement is zero at the ends of the string. This means that k must be such that the ends correspond to nodes of the wave. As shown in figure 2.4, we therefore have *modes of vibration*, in which there are

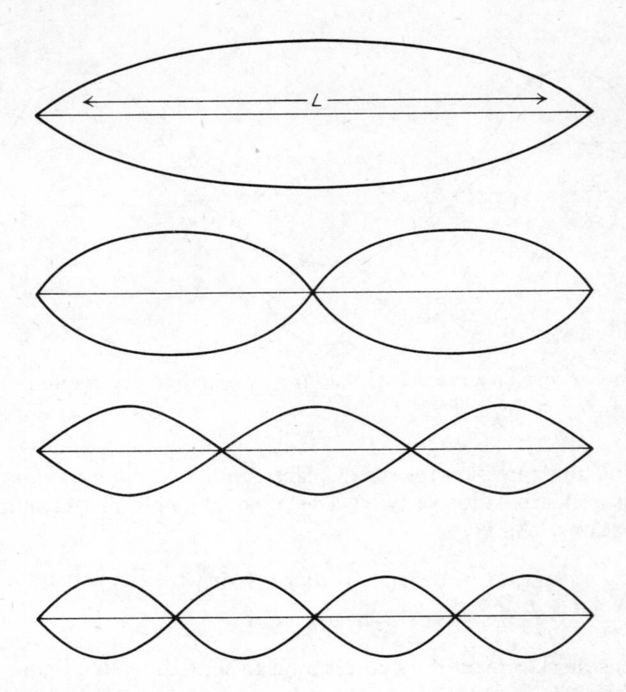

Figure 2.4 Successive standing wave modes on a string fixed at points L apart

successively 1, 2, 3 . . . antinodes between the ends of the string. Since the ends are nodes, the value of k must satisfy

$$k_N L = N\pi \tag{2.13}$$

where the *mode number* N is an integer, and in fact is equal to the number of antinodes on the wave. In terms of wavelength, N is the number of half wavelengths contained in the length L. Using equation 2.5 for the velocity v we may also convert equation 2.13 to an expression for the angular frequency ω_N of the Nth mode

$$\omega_N = N\pi v/L \tag{2.14}$$

It is also useful to have an expression for the separation in frequency $\delta\omega$ between successive modes

$$\delta\omega = (\omega_{N+1} - \omega_N) = \pi v/L \tag{2.15}$$

In fact $\delta\omega$ is equal to ω_1, the frequency of the $N = 1$, or *fundamental*, mode.

In most instances, standing waves are set up at a low mode number N. For example, a violin or guitar string is normally played at the fundamental, $N = 1$. The most important exception is the *laser*, or more exactly *laser oscillator*, which is used as an intense light source. A laser generates light in a very narrow band centred on its characteristic frequency, say ω_0, and when it is operating it contains a standing wave of light of frequency ω_0. The standing wave is usually set up

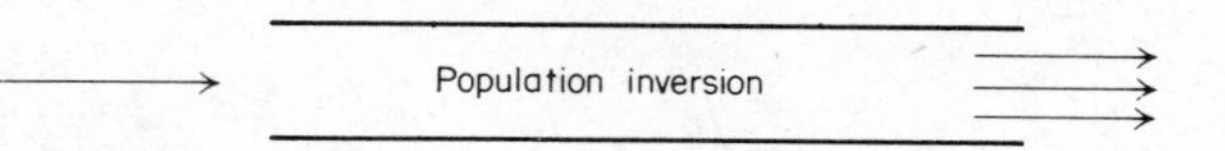

Figure 2.5 Light is amplified on passing through a gas in a state of population inversion

between two mirrors, one at each end of the laser. The distance between the mirrors is much greater than the optical wavelength, so the standing wave has a high mode number. In view of the importance of lasers as intense light sources, we conclude this section with a rather fuller account of laser operation.

The first essential constituent of a laser is a medium which amplifies light of frequency ω_0. In a *gas laser*, the frequency for amplification is that of a discharge line of the gas. For example, a He–Ne laser amplifies the familiar 633 nm red line of neon. The amplification mechanism is quantum mechanical, and need not concern us in any detail; it is sufficient to know that if the gas is brought into a

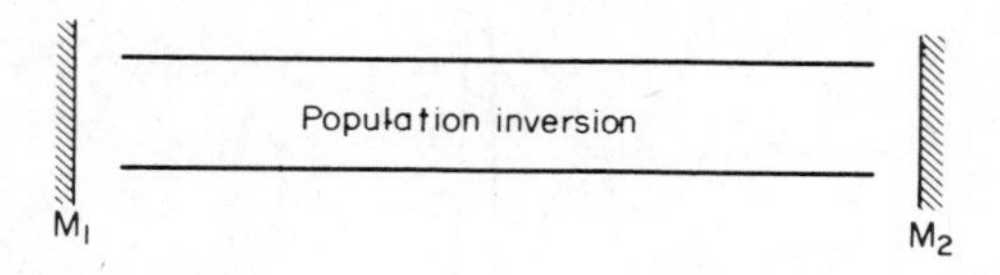

Figure 2.6 Laser oscillator consists of an amplifying medium, together with mirrors M_1 and M_2 to provide feedback

state of what is called *population inversion*, then light of the appropriate frequency is amplified (figure 2.5). Thus if light of a wavelength of about 633 nm passes through a column of neon in a state of population inversion, it emerges amplified. Many media can be brought into a state of population inversion and are therefore laser media. Besides a wide range of gases, laser media include various solids containing impurity ions. Two of the commonest solid laser media are ruby, which contains Cr^{3+} (triply ionised chromium) ions, and glass containing Nd^{3+} ions.

To turn a laser amplifier into an oscillator, or source of radiation, one must arrange for *feedback* of some of the amplified radiation. As stated, this is done simply by putting mirrors at each end of the amplifying medium, so that the radiation makes a number of successive passes through the medium (figure 2.6). A standing wave is set up between the mirrors and the space between the mirrors is called the *laser cavity*. If the laser is to be used as a source of light, some of the radiation must be allowed out, so one of the mirrors is made as a slightly imperfect reflector which transmits a small fraction, typically 1 per cent for a He - Ne laser, of the light falling on it. This is the light which we see emerging from the laser.

It is clear that the standing wave pattern in a typical laser corresponds to a very high mode number N, since the length of the laser is very much greater than the wavelength. For example, in a CO_2 laser of length 1 m, operating at a wavelength of 10.6 μm, N is of order 2×10^5. The fact that the laser operates at a high mode number has important practical consequences, since the mode separation $\delta\omega$ of equation 2.15 is very much smaller than the frequency of operation, equation 2.14. It will be recalled from section 1.6 that a discharge line in a gas does not in

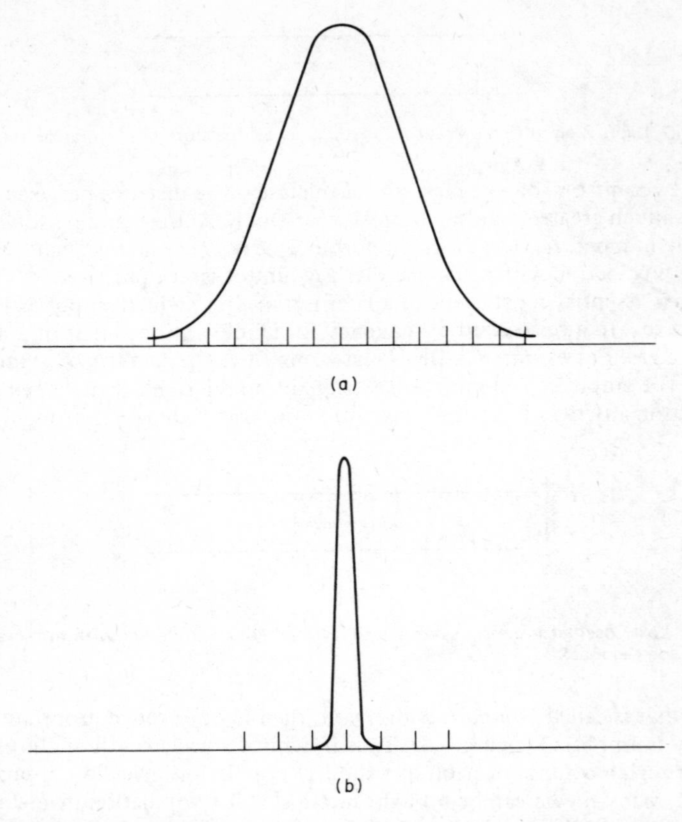

Figure 2.7 Two possibilities for laser operation. (a) Many modes within the linewidth. (b) Single mode within the linewidth

fact correspond to a single sharp frequency since in general it is both Doppler and collision broadened. Similarly, the line which is to be amplified in a solid laser is always broadened to some extent. Linewidths vary from around 100 MHz in a low-pressure CO_2 laser to as much as 4 THz (4000 GHz) in some neodymium-doped glass lasers. When the line broadening is substantial, there can be many standing wave modes within the linewidth, as sketched in figure 2.7(a). In this case, the laser can in principle operate in any of the modes which fall within the linewidth, and unless special precautions are taken the frequency of oscillation jumps fairly rapidly between the modes. This phenomenon of mode jumping means that a laser of this kind emits light of the same bandwidth as the line which is being amplified, so that the laser light is no more coherent than light from an ordinary source using that line. On the other hand, it is possible, for example in a low-pressure gas laser, for there to be only one mode within the linewidth, as sketched in figure 2.7(b). This second kind of laser oscillates in the single mode, and thus can be a highly coherent source. However, it is important to realise that by no means all lasers are sources of coherent light.

Worked example 2.1

Calculate the fundamental frequency and mode separation of standing waves on a string stretched between two points 0.3 m apart, at a tension such that the velocity of sound on the string is 3×10^2 m s^{-1}.

Answer

From equations 2.14 and 2.15, the fundamental frequency is

$$\omega_1 = \pi v/L$$

and the mode separation $\delta\omega$ is equal to the fundamental frequency. Putting in the numerical values we find

$$\omega_1 = \frac{\pi \times 3 \times 10^2}{3 \times 10^{-1}} = 10^3 \pi \text{ s}^{-1}$$

The frequency in hertz is

$$\delta f = f_1 = \frac{\omega_1}{2\pi} = 500 \text{ Hz}$$

2.3 Longitudinal and Transverse Waves. Polarisation

We now come to a basic distinction between types of waves, which we can illustrate first of all with the two main types of acoustic waves. A sound wave in air, figure 2.8, consists of successive regions of compression and rarefaction. The

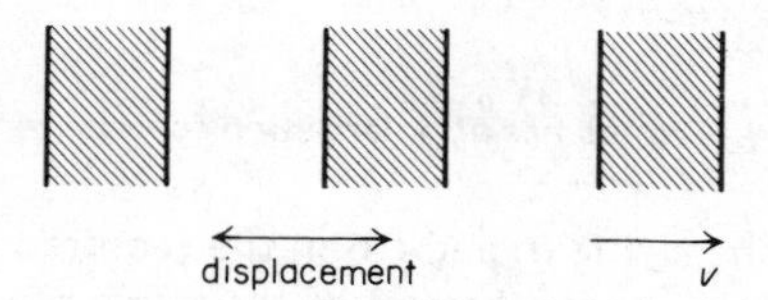

Figure 2.8 Sound wave in air

displacement is in the same direction as the velocity, so we have a *longitudinal wave.* On the other hand, in a wave on a string (figure 2.9), the displacement is at right angles to the velocity, so we have a *transverse wave.* It should be pointed out straight away that a wave need not be either purely longitudinal or purely transverse; for example in a surface wave on water the particles move in the approximately circular paths shown in figure 2.10. However, most of the waves we shall deal with are either longitudinal or transverse. A transverse wave is more involved

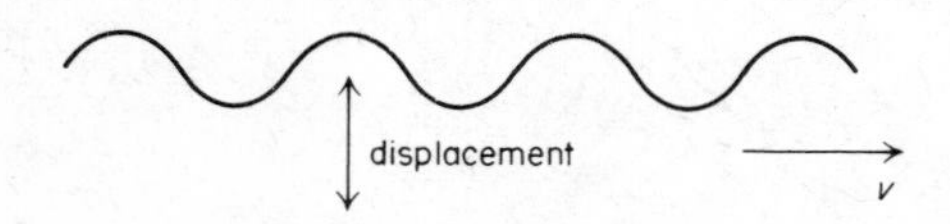

Figure 2.9 Wave on a string

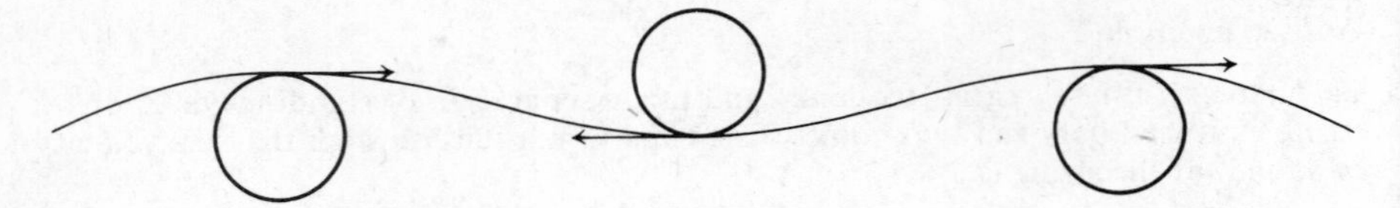

Figure 2.10 Particle paths in surface wave on water.

than a longitudinal wave, because the displacement can be in any direction in the plane at right angles to the velocity. For example, the wave on a string can have the displacement 'vertical' or 'horizontal', as in figure 2.11. If the displacement is always in one direction, as in either wave in figure 2.11, the wave is called *plane polarised.* However, it is possible to generate more complicated polarisation states. The plane polarised waves of figure 2.11 are generated, obviously, if

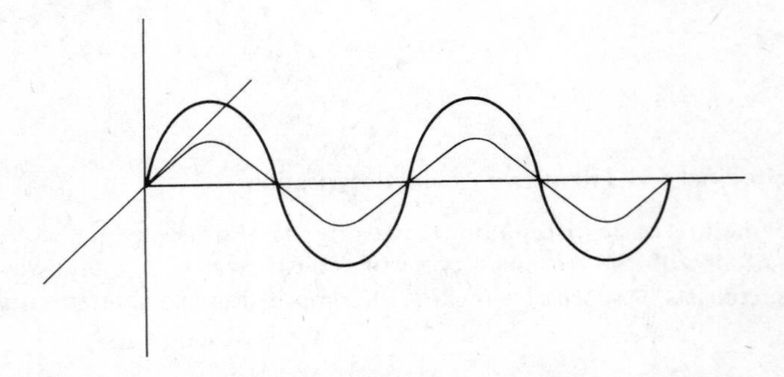

Figure 2.11 Two planes of polarisation for a wave on a string

one moves the end of the string to and fro in the required plane at the appropriate frequency. If, instead, one moves the end of the string round in a circle at the same frequency, then what travels down the string is a *circularly polarised* wave. The displacement at any point of the string is now not confined to one direction; instead it moves round in a circle.

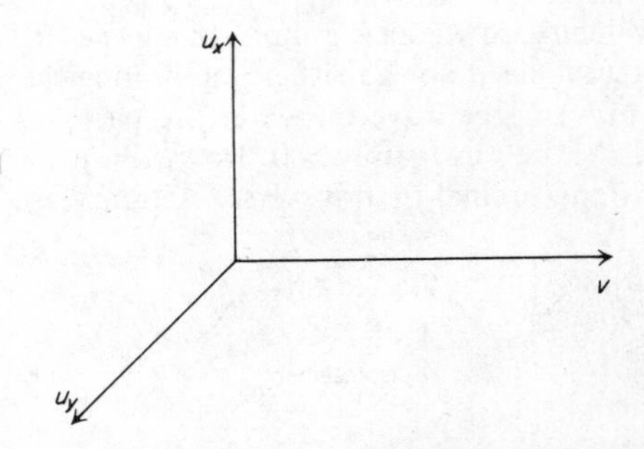

Figure 2.12 Notation for general polarisation states

We may discuss general polarisation states by specifying the displacements in the two directions at right angles to the velocity. Choose the z axis to lie along the velocity, so that we have to deal with the displacements u_x and u_y in the x and y directions (figure 2.12). In the next section we shall come across a special type of crystalline material, called birefringent, in which u_x and u_y propagate with different velocities. However in most ordinary materials, such as glass, both u_x and u_y propagate in the z direction with the same velocity v. It is then sufficient to study their amplitude and relative phase at one point z; we may use the point $z = 0$, so that the value of z does not appear explicitly. Since we are restricting our attention to a single frequency ω, u_x and u_y both oscillate at this frequency. If one or other displacement has zero amplitude, we have plane polarisation

$$\begin{aligned} u_x &= A\cos(\omega t + \eta) \qquad \text{plane polarised in } x \text{ direction} \\ u_y &= 0 \end{aligned} \tag{2.16}$$

$$\begin{aligned} u_x &= 0 \qquad \text{plane polarised in } y \text{ direction} \\ u_y &= A\cos(\omega t + \eta) \end{aligned} \tag{2.17}$$

For polarisation at a general angle to the x axis, we take appropriate amplitudes with the displacement in phase

$$\begin{aligned} u_x &= A\cos(\omega t + \eta) \qquad \text{plane polarised at angle } \theta \text{ to } x \text{ axis} \\ u_y &= B\cos(\omega t + \eta) \end{aligned} \tag{2.18}$$

It can be seen from figure 2.13 that the plane of polarisation makes an angle $\theta = \tan^{-1} B/A$ with the x axis.

To see the effect of a relative phase difference between u_x and u_y, we may look at the simplest case of equal amplitude and a relative phase of $\pi/2$

$$\begin{aligned} u_x &= A\cos(\omega t + \eta) \\ u_y &= A\sin(\omega t + \eta) \end{aligned} \tag{2.19}$$

It is clear that the displacement point (u_x, u_y) now moves around in the $x - y$ plane; for instance at $(\omega t + \eta) = 0$, $u_y = 0$ and $u_x \neq 0$, while at $(\omega t + \eta) = \pi/2$,

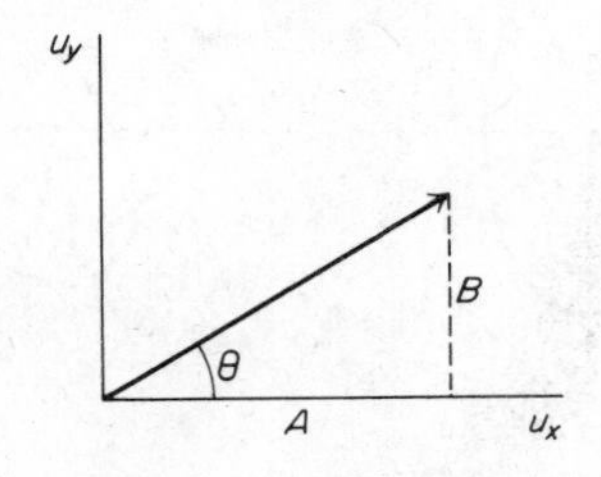

Figure 2.13 Displacement diagram for general linear polarisation

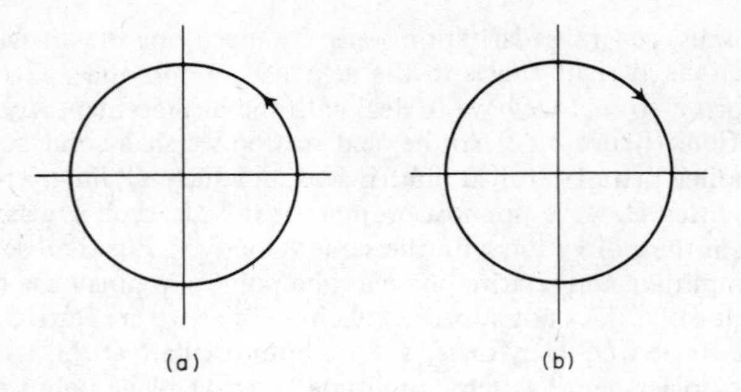

Figure 2.14 (a) Left-hand and (b) right-hand circular polarisation, corresponding to equations 2.19 and 2,21 respectively

$u_x = 0$ and $u_y \neq 0$. We can eliminate t to find how the point (u_x, u_y) moves: if we square and add the two parts of equation 2.19 we find

$$u_x{}^2 + u_y{}^2 = A^2 \tag{2.20}$$

Thus the point moves around a circle of radius A, as in figure 2.14(a). This is the circular polarisation which one generates by moving the end of a string in a circle. It is easy to see that the displacement point in figure 2.14(a) moves anticlockwise; for at $t = 0$, $u_x = A$ and $u_y = 0$, while at a slightly later time u_x is slightly decreased and u_y is small and positive. This state is called *left-hand circular polarisation. Right-hand circular polarisation* (figure 2.14b) is produced if the relative phase is $-\pi/2$ rather than $+\pi/2$:

$$\begin{aligned} u_x &= A \cos(\omega t + \eta) \\ u_y &= -A \sin(\omega t + \eta) \end{aligned} \tag{2.21}$$

With unequal amplitudes the general state is *elliptical polarisation.* For example, if

$$\begin{aligned} u_x &= A \cos(\omega t + \eta) \\ u_y &= B \sin(\omega t + \eta) \end{aligned} \tag{2.22}$$

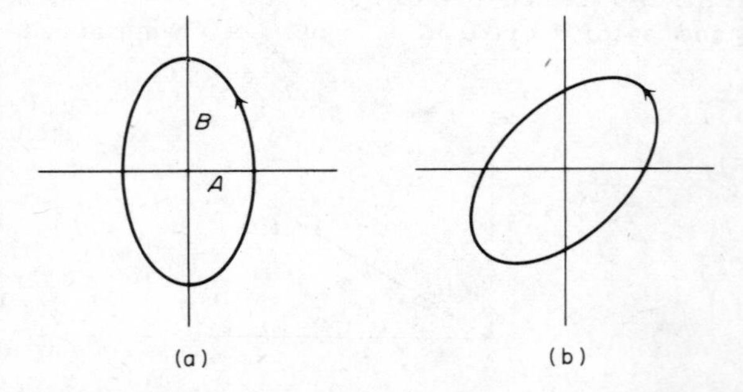

Figure 2.15 Elliptical polarisation (a) corresponding to equation 2.23 and (b) general.

then

$$\frac{u_x^2}{A^2} + \frac{u_y^2}{B^2} = 1 \tag{2.23}$$

and the polarisation point (u_x, u_y) moves round an ellipse whose principal axes are in the x and y directions (figure 2.15). Finally, if we have arbitrary relative phase and unequal amplitudes, the point (u_x, u_y) moves round an ellipse whose principal axes do not coincide with the x and y axes (problem 2). Thus elliptical polarisation is the most general polarisation state.

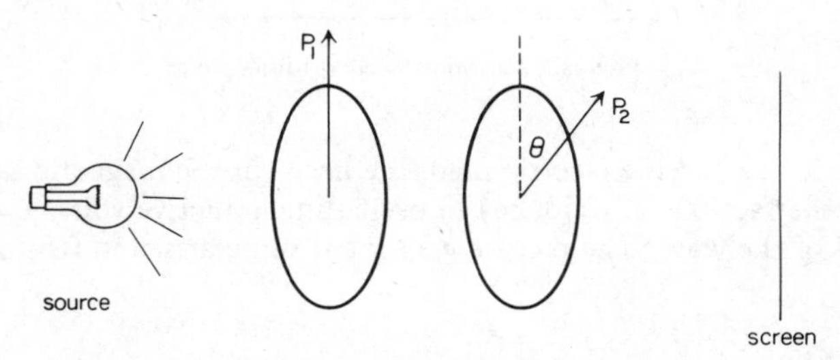

Figure 2.16 Transmission of light through two sheets of polaroid. When θ, the angle between the preferred axes P_1 and P_2, is 90°, no light passes

We have written the displacements u_x and u_y with a general phase η, as in equation 2.18 and 2.19. However, η itself does not affect the polarisation state; what matters is the *relative phase* of u_x and u_y. Thus if the relative phase is zero, as in equation 2.18, we always find plane polarisation. If the relative phase is $\pi/2$, as in equation 2.19 and 2.22, we find circular or elliptical polarisation. In the next section, we discuss the use of birefringent material to construct *retardation plates* which are devices for manipulating the relative phase.

Electromagnetic waves, typically light, are the most important kind of transverse radiation. That they are transverse follows from the fact that they can be polarised. If one places a sheet of Polaroid in front of a light source, and then rotates a second sheet of Polaroid between the first sheet and a screen (figure 2.16), the light from the source is completely extinguished at one particular orientation of the second sheet. At 90° from the extinction point, all the light passing through the first sheet of Polaroid also passes through the second. The reason is that Polaroid has a *preferred axis*, and light incident upon it emerges plane polarised with the displacement in the direction of the preferred axis. When the axes of the two sheets are at right angles, all the light passing through the first sheet is cut out by the second.

To understand how Polaroid works, we can turn to a *microwave grid polariser.* Microwaves are electromagnetic radiation of wavelengths between roughly 1 mm and 10 cm, and they can be polarised with the grid sketched in figure 2.17, that is, a series of parallel metal rods or wires. It was mentioned in chapter 1 that the displacement in an electromagnetic wave is an oscillating electric field or E field; we now know that the wave is transverse, so that the E field is at right angles to the velocity. In addition, there is an oscillating magnetic field at right angles to

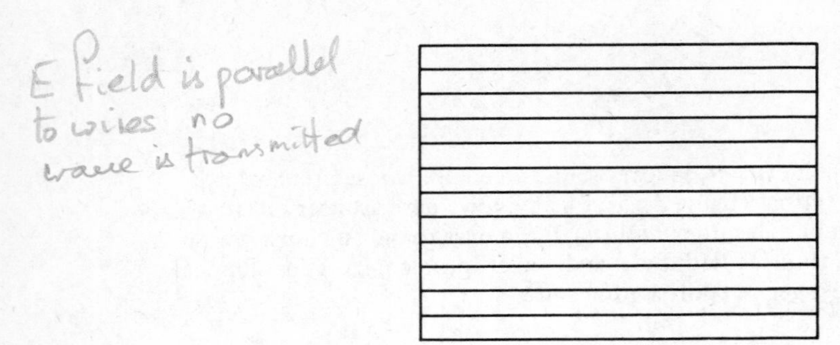

Figure 2.17 Microwave grid polariser

both the velocity and the electric field; the need for the magnetic field can be seen from the fact that it produces an oscillating inductive voltage, which is just the E field of the wave. The picture we have is summarised in figure 2.18. Now

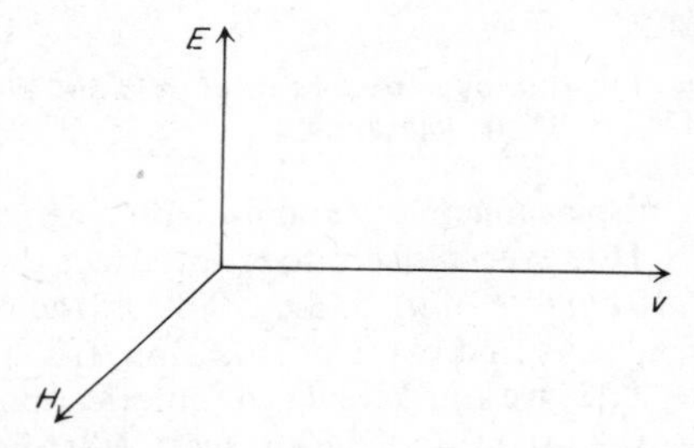

Figure 2.18 Velocity, electric field and magnetic field are mutually at right angles in an electromagnetic wave

when microwaves fall on a grid polariser, what happens depends on the orientation of the E field relative to the metal bars. If the E field is along the bars, electric currents are set up in the bars, energy is removed, and the wave does not pass the grid. If the E field is perpendicular to the bars, the wave is transmitted, unaffected. Thus if the incident radiation is composed of a mixture of polarisation states, the transmitted radiation is plane polarised with the E field at right angles to the bars, so that the grid acts as a polariser. Polaroid is analogous to the grid polariser, in that it consists of long organic molecules, all aligned in one particular direction. The structure is of course on a much finer scale, so that Polaroid is effective in polarising light, which has a much shorter wavelength than microwaves.

We should emphasise finally that throughout this section we have been discussing specific polarisation states. As we have mentioned already, in section 1.5, the light emitted by a single atom in a hot gas is a succession of distinct pulses, and the light emitted from all the atoms in a discharge tube is a random sequence of such pulses. Each pulse may have its polarisation state, but the light as a whole is *unpolarised;* that is, it is not in any specific polarisation state.

2.4 Retardation Plates

As we have seen, plane polarised light can be produced with the aid of a polariser. It can then be manipulated, and converted into other polarisation states, with a class of components known as *retardation plates.* Certain crystals, for example mica, have inherent x and y axes such that there are two velocities of propagation, one for light with the plane of polarisation along the crystal x axis, and another for light with the plane of polarisation along the crystal y axis. This property is called *birefringence*. Since we are dealing with light of one frequency ω, we must now use different wave numbers $k = \omega/v$ for displacements in the x and y directions, so that we write a running wave in a birefringent crystal as

$$u_x = A\cos(\omega t - k_1 z - \eta_1)$$
$$u_y = B\cos(\omega t - k_2 z - \eta_2) \qquad (2.24)$$

Now suppose a wave is incident upon a slab of birefringent material extending from $z = 0$ to $z = z_0$ (figure 2.19). We take equal amplitudes for simplicity, and write the incident wave as

$$u_x = A\cos(\omega t - \eta_1)$$
$$u_y = A\cos(\omega t - \eta_2) \qquad (2.25)$$

The relative phase is $(\eta_1 - \eta_2)$, so that, for example, if $(\eta_1 - \eta_2) = 0$ we have linear polarisation, and if $(\eta_1 - \eta_2) = \pi/2$ we have circular polarisation. We can see from equation 2.24 that after transmission through the crystal, the wave has the form

$$u_x = A\cos(\omega t - \eta_1 - k_1 z_0)$$
$$u_y = A\cos(\omega t - \eta_2 - k_2 z_0) \qquad (2.26)$$

The relative phase is now $(\eta_1 - \eta_2) + (k_1 - k_2)z_0$, and so the relative phase has shifted by

$$\eta_r = (k_1 - k_2)z_0 \qquad (2.27)$$

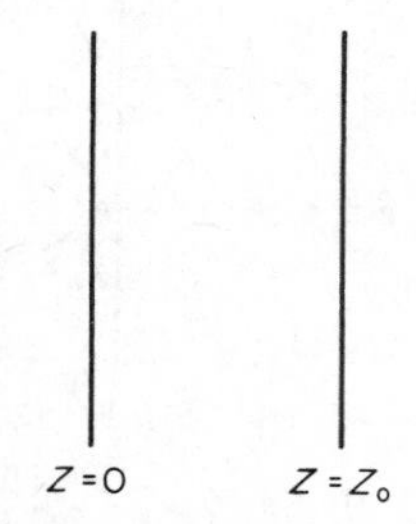

Figure 2.19 Birefringent slab

A slab of birefringent material such as we have described is called a retardation plate and it is characterised by the phase change η_r. The two examples commonly met are

$$\eta_r = \pi/2 \qquad \text{quarter-wave plate} \tag{2.28}$$

$$\eta_r = \pi \qquad \text{half-wave plate} \tag{2.29}$$

Note that the retardation plate has its own x and y axes, and the consequences of passing plane polarised light through a retardation plate depend on the angle between the plane of polarisation and the axes of the retardation plate. For example, if we pass plane polarised light through a quarter-wave plate, we may have

Before	After	
$u_x = A \cos \omega t$	$u_x = A \cos(\omega t - \eta_x)$	
$u_y = 0$	$u_y = 0$	No change

if the plane of polarisation of the incident light is along the x axis of the quarter-wave plate. We write $\eta_x = k_1 z_0$ for the change in phase of u_x; the emergent light still has the same plane of polarisation, so that the plate produces no change. Similarly, if the incident light has the plane of polarisation along the y axis of the quarter-wave plate, we find

Before	After	
$u_x = 0$	$u_x = 0$	
$u_y = A \cos \omega t$	$u_y = A \cos(\omega t - \eta_y)$	No change

However, if the plane of polarisation of the incident light is at 45° to the axes, we find

Before	After	
$u_x = A \cos \omega t$	$u_x = A \cos(\omega t - \eta_x)$	
$u_y = A \cos \omega t$	$u_y = A \cos(\omega t - \eta_y)$	
	$= A \cos(\omega t - \eta_x + \pi/2)$	Conversion from linear to circular
	$= -A \sin(\omega t - \eta_x)$	

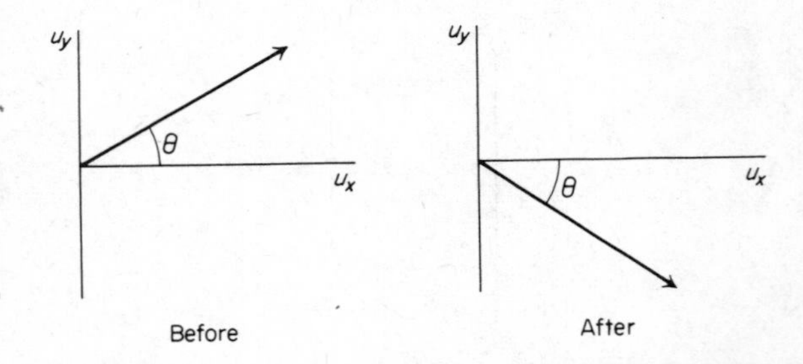

Figure 2.20 Rotation of plane of polarisation by a half-wave plate

Conversely, a quarter-wave plate always converts circular polarisation to linear.

A half-waveplate reverses the sign of u_y relative to u_x. Thus it rotates the plane of polarisation

Before	After
$u_x = A \cos \omega t$	$u_x = A \cos(\omega t - \eta_x)$
$u_y = B \cos \omega t$	$u_y = -B \cos(\omega t - \eta_x)$

As shown in figure 2.20, this corresponds to a rotation of the plane of polarisation through the angle 2θ, from θ above the x axis to θ below, where $\theta = \tan^{-1}(B/A)$ is the angle marked in figure 2.13. It can be seen similarly that a half-wave plate converts circular polarisation from left hand to right hand and vice versa.

Worked example 2.2

Find the effect of passing elliptically polarised light through a half-wave plate whose axes coincide with the principal axes of the ellipse. What happens if the half-wave plate is replaced by a quarter-wave plate?

Answer

The elliptically polarised light may be written as in equation 2.22

$$u_x = A \cos \omega t$$
$$u_y = B \sin \omega t$$

where the x and y axes are the axes of the half-wave plate. As shown in figure 2.15(a), this is left-hand elliptical polarisation. After passing through the half-wave plate, the light becomes

$$u_x = A \cos \omega t$$
$$u_y = -B \sin \omega t$$

We still have

$$\frac{u_x^2}{A^2} + \frac{u_y^2}{B^2} = 1$$

so the light remains elliptically polarised. However, the polarisation point (u_x, u_y) now moves round the ellipse in the opposite sense, as can be seen by considering the motion from time $t = 0$. We now have right-hand elliptical polarisation.

If the half-wave plate is replaced by a quarter-wave plate, the emergent light becomes

$$u_x = A \cos \omega t$$
$$u_y = B \cos \omega t$$

which as in figure 2.13 corresponds to linear polarisation at an angle $\theta = \tan^{-1} B/A$ to the x axis.

2.5 Propagation of General Wave Forms

We saw in section 1.4 that we may regard a general waveform as a superposition of sinusoidal waves of different frequencies. In order to discuss the propagation of a general waveform, therefore, we must start by considering how the propagation of a single sinusoidal wave depends on its frequency. The crucial point is that, in

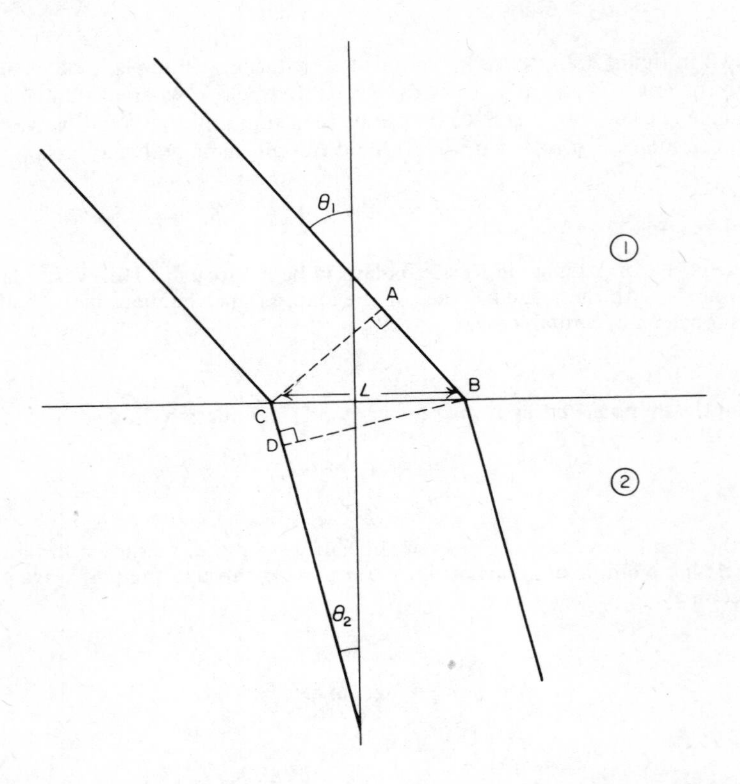

Figure 2.21 To illustrate the derivation of Snell's law for light passing from medium 1 to medium 2

general, the velocity of propagation is a function of the frequency. Let us first recall the derivation of Snell's law for the refraction of light passing from one medium to another, say from air to glass. We consider a portion of the incident wavefront which intercepts a length L on the plane interface between the media (figure 2.21). The wave in medium 1 travels a distance AB while the wave in medium 2 travels a distance CD. Equating the corresponding times, we find

$$\frac{L \sin \theta_1}{v_1} = \frac{L \sin \theta_2}{v_2} \tag{2.30}$$

where v_1 and v_2 are the velocities in the two media. Cancelling the factor L from either side, we get

$$\frac{\sin \theta_1}{v_1} = \frac{\sin \theta_2}{v_2} \tag{2.31}$$

which is Snell's law. Frequently, the velocities are written as $v = c/n$ where c is the velocity of light in a vacuum and n is the *refractive index,* so that Snell's law takes the form

$$n_1 \sin \theta_1' = n_2 \sin \theta_2 \tag{2.32}$$

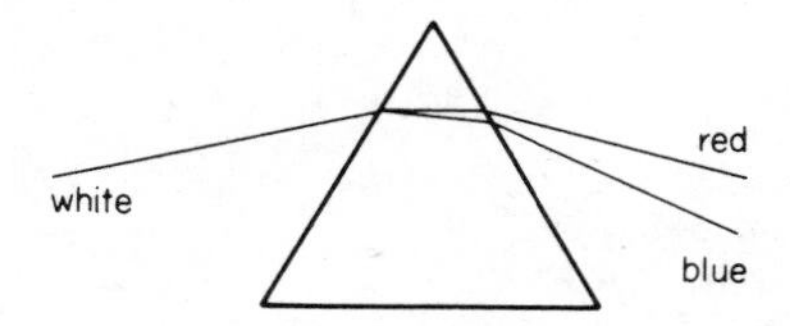

Figure 2.22 Refraction of white light by a prism

For our purposes, it is sufficient to note that refraction depends on light having different velocities in the two media.

The breaking up of white light by a prism (figure 2.22) is now sufficient to show us that the velocity of light in glass depends on the frequency; in fact blue light is refracted more strongly (blue bends best), so we conclude that blue light has a lower velocity than red light in glass.

A medium in which velocity depends on frequency is called *dispersive;* the term is used because dispersion of light by a prism is the best known example. It is convenient to characterise a dispersive medium, not by quoting velocity as a function of frequency, but rather by giving angular frequency ω as a function of angular wave number k. We shall see that the latter is more useful, and since $v = \omega/k$ we could, if necessary, obtain $v(\omega)$ from the ω-k plot. The ω-k relation is sometimes called the *dispersion equation.* By way of example, we may take the propagation of longitudinal waves on a monatomic lattice such as a system of equal masses linked by springs (figure 2.23). This system, of course, serves as a model for longitudinal sound waves in a crystal. The relationship between ω and k for this system is

$$\omega = 2\sqrt{\left(\frac{C}{\rho}\right)\frac{1}{a}}\sin\frac{ka}{2} \tag{2.33}$$

where $\rho = m/a$ is the average density, and C is an elastic constant, related to the spring constant. The graph of equation 2.33 is shown in figure 2.24. We may note

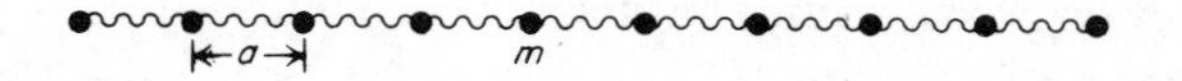

Figure 2.23 System of masses linked by springs

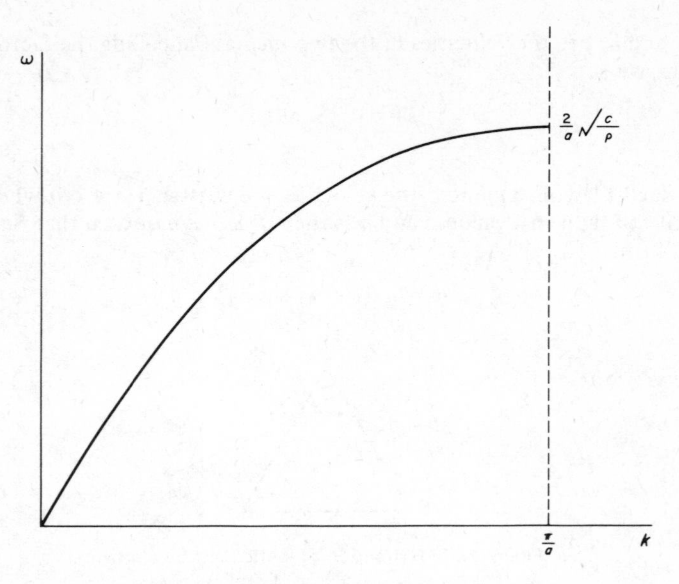

Figure 2.24 $\omega - k$ plot for system of figure 2.23

that for small values of k, equation 2.33 simplifies, since we can replace $\sin \frac{1}{2}ka$ by $\frac{1}{2}ka$

$$\omega = vk \qquad \text{for } k \ll \pi/a \tag{2.34}$$

with

$$v = \sqrt{\frac{C}{\rho}} \tag{2.35}$$

This corresponds to the straight line region near the origin in figure 2.24. In this region, the velocity ω/k is independent of k and we have *non-dispersive* propagation. The reason is straightforward. Small k corresponds to long wavelength, and when the wavelength is much greater than the spacing between the masses in figure 2.23, the medium simply appears to the wave as an elastic continuum. The wave then propagates as ordinary long-wavelength sound. We shall see in chapter 6 that equation 2.35 is indeed the velocity of long-wavelength sound. As k increases, the wavelength decreases, until the particle structure of the medium begins to affect matters, and the propagation becomes dispersive.

We are now ready to deal with the topic we introduced at the beginning of this section, namely the propagation of a waveform of general shape. Suppose at time $t = 0$, we have a wave $u(x)$ in a dispersive medium (figure 2.25). We may write $u(x)$ as a sum of sinusoidal waves of different wave numbers

$$u(x) = \sum_n u_n \cos(k_n x + \delta_n) \tag{2.36}$$

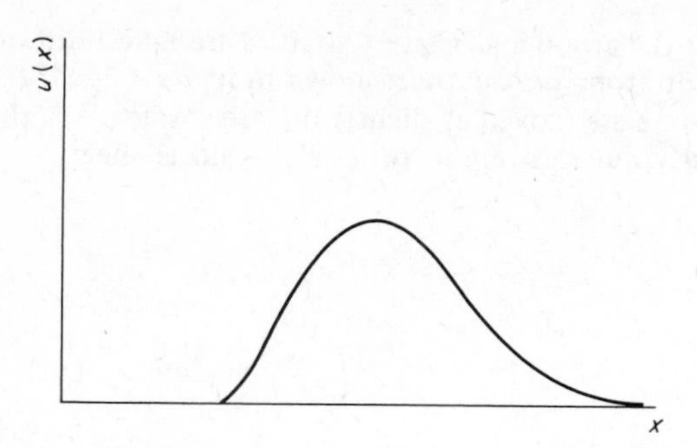

Figure 2.25 General waveform $u(x)$

Now each sinusoidal wave travels with its own velocity $v(k)$ so at some later time t each has travelled a different distance $v(k)t$ and the phase relations between the different waves have changed. Consequently the sinusoidal waves sum to a function $u_1(x, t)$ which does not have the same shape as $u(x)$. This shows that the first property of a dispersive medium is that a general disturbance changes in shape as it propagates. We can find the second, and perhaps more important, property if we look at the propagation of the simplest kind of superposition, which is a beat pattern.

Let us add two travelling sine waves, of equal amplitude for simplicity

$$\begin{aligned} u &= u_0 \cos(k_1 x - \omega_1 t) + u_0 \cos(k_2 x - \omega_2 t) \\ &= 2u_0 \cos[\tfrac{1}{2}(k_1 - k_2)x - \tfrac{1}{2}(\omega_1 - \omega_2)t] \cos[\tfrac{1}{2}(k_1 + k_2)x - \tfrac{1}{2}(\omega_1 + \omega_2)t] \end{aligned} \tag{2.37}$$

The first term is the beat envelope, and the second the carrier, and at any one instant of time the waveform looks like the beat pattern shown in figure 1.9. However, we see from equation 2.37 that the envelope and the carrier travel at different velocities

$$v_{\text{beat}} = (\omega_1 - \omega_2)/(k_1 - k_2) \tag{2.38}$$

$$v_{\text{carrier}} = (\omega_1 + \omega_2)/(k_1 + k_2) \tag{2.39}$$

These velocities are the same and equal to v in a non-dispersive medium, for which $\omega = vk$, but they differ in a dispersive medium. As in section 1.3 we shall concentrate on the situation when ω_1 and ω_2 are nearly equal, when we find

$$v_{\text{beat}} = \frac{\mathrm{d}\omega}{\mathrm{d}k} = v_{\mathrm{g}} \tag{2.40}$$

$$v_{\text{carrier}} = \frac{\omega}{k} = v_{\mathrm{p}} \tag{2.41}$$

where we now introduce the terminology *group velocity* v_{g} and *phase velocity* v_{p} for the velocities of the beat and carrier respectively.

The fact that the group and phase velocities are indeed different can readily be seen with the Letratone beat pattern shown in figure 1.12. If the two superposed sheets of Letratone are moved at slightly different velocities, the wave groups move at a velocity quite different from that of either sheet.

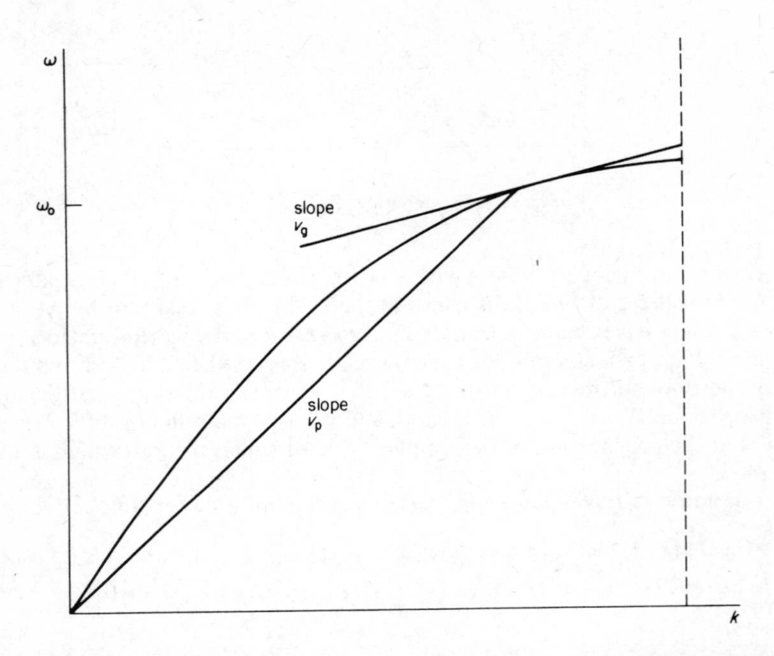

Figure 2.26 Calculation of group velocity v_g and phase velocity v_p for longitudinal waves on a monatomic lattice

We may illustrate the calculation of group and phase velocity with equation 2.33 for the dispersion of longitudinal waves on a monatomic lattice. As sketched in figure 2.26 the phase velocity at a frequency ω_0 is the slope of the chord to that point on the dispersion curve, and the group velocity is the slope of the tangent. Explicitly, equations 2.40 and 2.41 give

$$v_g = \sqrt{\left(\frac{C}{\rho}\right)} \cos \frac{ka}{2} \tag{2.42}$$

$$v_p = \sqrt{\left(\frac{C}{\rho}\right)} \frac{2}{ka} \sin \frac{ka}{2} \tag{2.43}$$

which are sketched in figure 2.27. The group and phase velocities are both equal to $\sqrt{(C/\rho)}$ for small k, the region of non-dispersive propagation, and thereafter v_g falls below v_p.

We can now qualify our earlier statement that a general waveform changes shape as it travels through a dispersive medium. If the waveform contains only a narrow range of frequencies, then it travels an appreciable distance without change of shape, and it travels at the group velocity. An example is given in figure 2.28

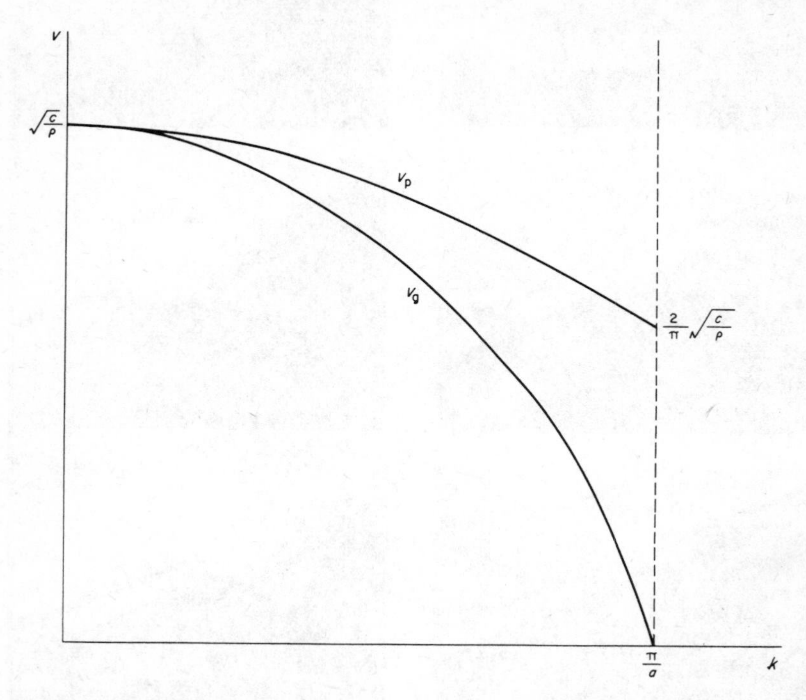

Figure 2.27 Group and phase velocities as functions of k for longitudinal waves on a monatomic lattice

which shows stills from a film of the propagation of a wave packet on the surface of water. Here the group velocity is smaller than the phase velocity and it can be seen that the individual wave crests do move up through the main wave packet.

Worked example 2.3

The *whistler mode* is a wave which travels along magnetic field lines in a plasma. It has the dispersion relation

$$\frac{k^2c^2}{\omega^2} = \frac{\omega_p^2/\omega^2}{1 - \omega/\omega_c}$$

where ω_p and ω_c are two fundamental frequencies (the plasma frequency and the cyclotron resonance frequency respectively) which satisfy $\omega_p \gg \omega_c$. Sketch the dispersion curve, and find the group and phase velocities as functions of frequency ω.

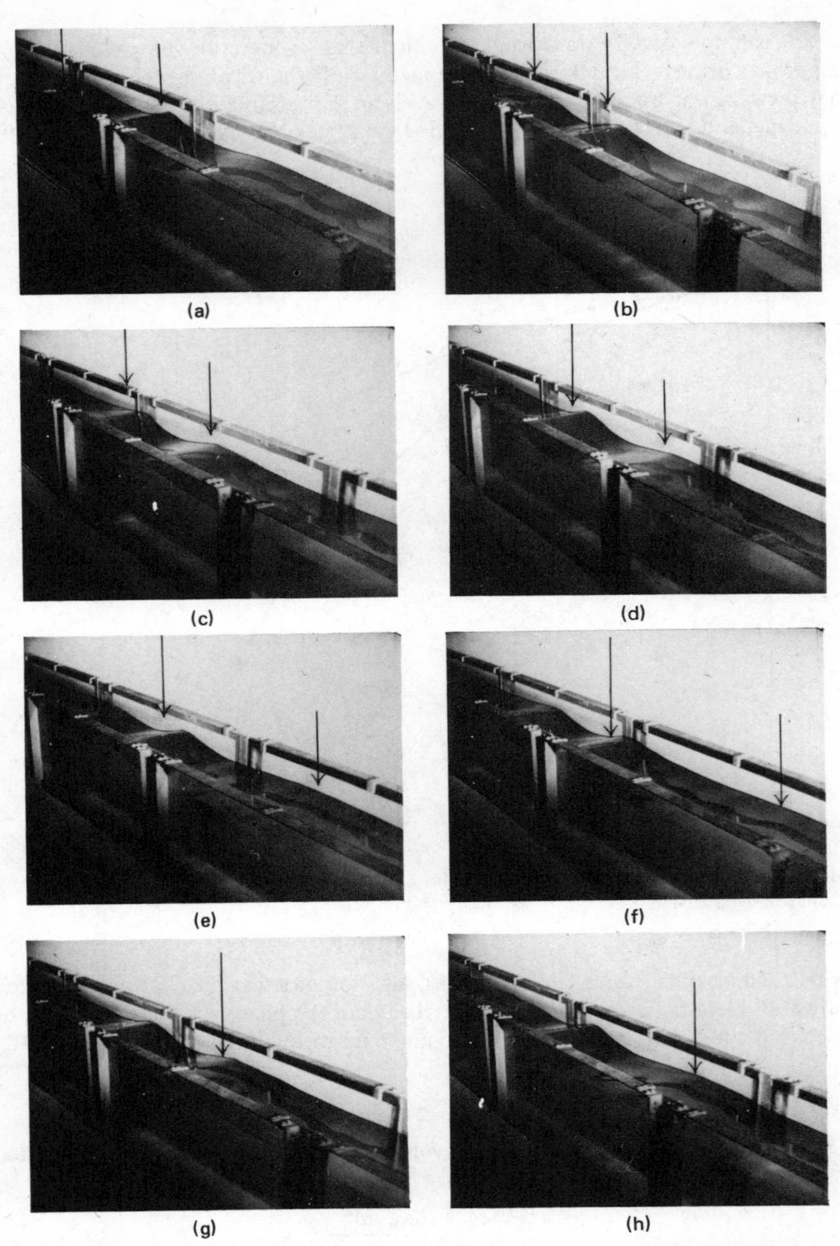

(a) (b) (c) (d) (e) (f) (g) (h)

Figure 2.28 Successive stages in the motion of a wave packet along a water trough. The photographs were taken from a camera moving at the group velocity, v_g. The motion of two particular crests is singled out. (Courtesy of Dr G. Bekow. From the film by E. David and G. Bekow, ***Gruppen-und Phasengeschwindigkeit (Group and Phase Velocity),*** Film C 614 of the Institut für den Wissenschaftlichen Film, Göttingen, (West Germany (1953))

Answer

To find the dispersion curve, we regard k as a function of ω

$$k = \frac{\omega_p/c}{(1 - \omega/\omega_c)^{1/2}}$$

This has the finite value $k = \omega_p/c$ at $\omega = 0$, and $k \to \infty$ as $\omega \to \omega_c$. The dispersion curve is therefore as sketched below

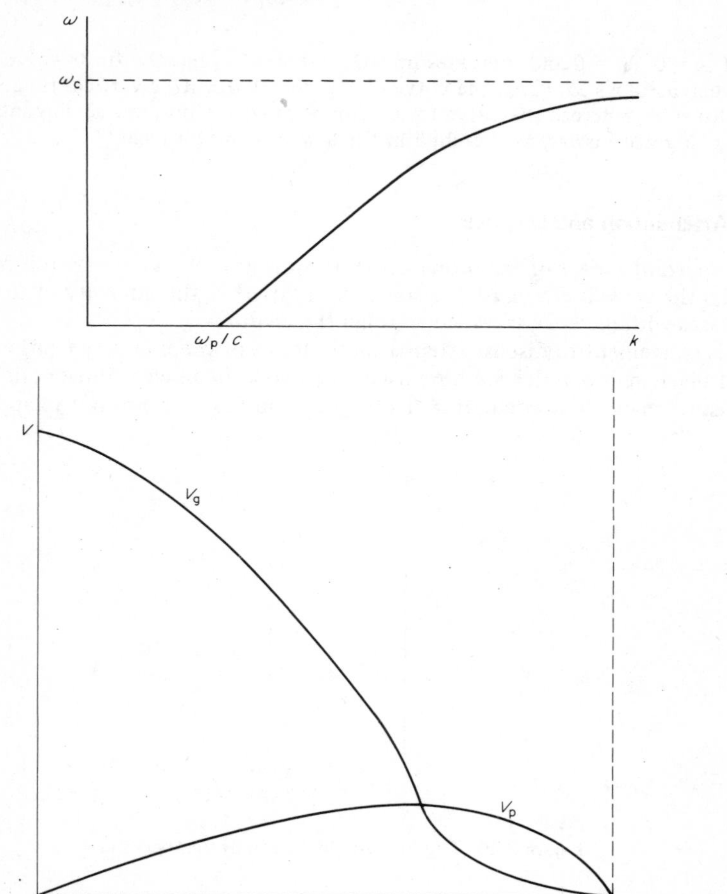

It is convenient to plot v_p and v_g as functions of ω over the frequency range of interest, namely from 0 to ω_c. The phase velocity is

$$v_p = \frac{\omega}{k} = \frac{\omega(1 - \omega/\omega_c)^{1/2}}{\omega_p/c}$$

The group velocity is

$$v_g = \frac{d\omega}{dk} = 1/(dk/d\omega)$$

and since

$$\frac{dk}{d\omega} = \frac{\omega_p / 2c\omega_c}{(1 - \omega/\omega_c)^{3/2}}$$

we have

$$v_g = \frac{2c\omega_c}{\omega_p} (1 - \omega/\omega_c)^{3/2}$$

At $\omega = 0$, $v_p = 0$ and increases linearly, whereas v_g has the finite value $2c\omega_c/\omega_p$ and decreases. As $\omega \to \omega_c$, the curve for v_p comes in with a vertical tangent, $dv_p/d\omega \to \infty$, whereas the curve for v_g comes in with a horizontal tangent. The curves therefore cross, as sketched in the lower figure on page 55.

2.6 Attenuation and Decibels

An important aspect of the propagation of a wave through a medium is that in general the wave is *attenuated* to some extent; that is, the intensity of the wave decreases with distance travelled through the medium.

It is convenient to discuss attenuation in terms of the intensity I rather than the displacement u which we have used up to now. In an electromagnetic wave, the displacement is an oscillating electric field, and as we point out in appendix 1,

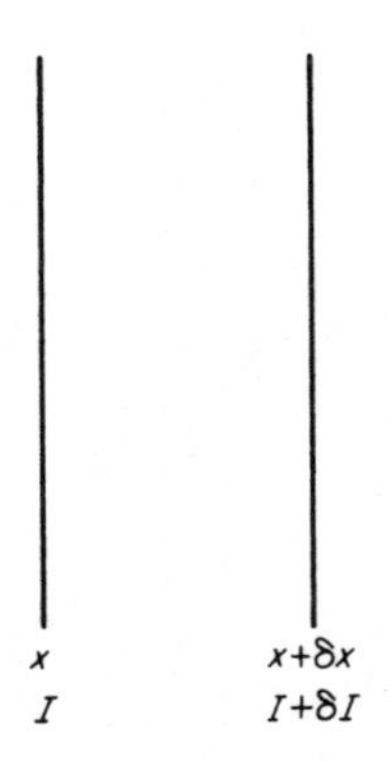

Figure 2.29 To illustrate discussion of Lambert's law

the intensity, which has units of watts per square metre, is proportional to the time average of the square of the displacement

$$I \propto \langle u^2 \rangle \tag{2.44}$$

We shall see in chapter 6 that equation 2.44 holds for sound waves as well.

Usually, in an attenuating medium, the intensity falls off exponentially with distance travelled through the medium

$$I = I_0 \exp(-\lambda x) \tag{2.45}$$

where the x axis is in the direction of travel of the wave. Equation 2.45 is known as *Lambert's law*; we can derive it, or at least make it plausible, in the following manner. Suppose that the wave has intensity I at point x and $I + \delta I$ at $x + \delta x$

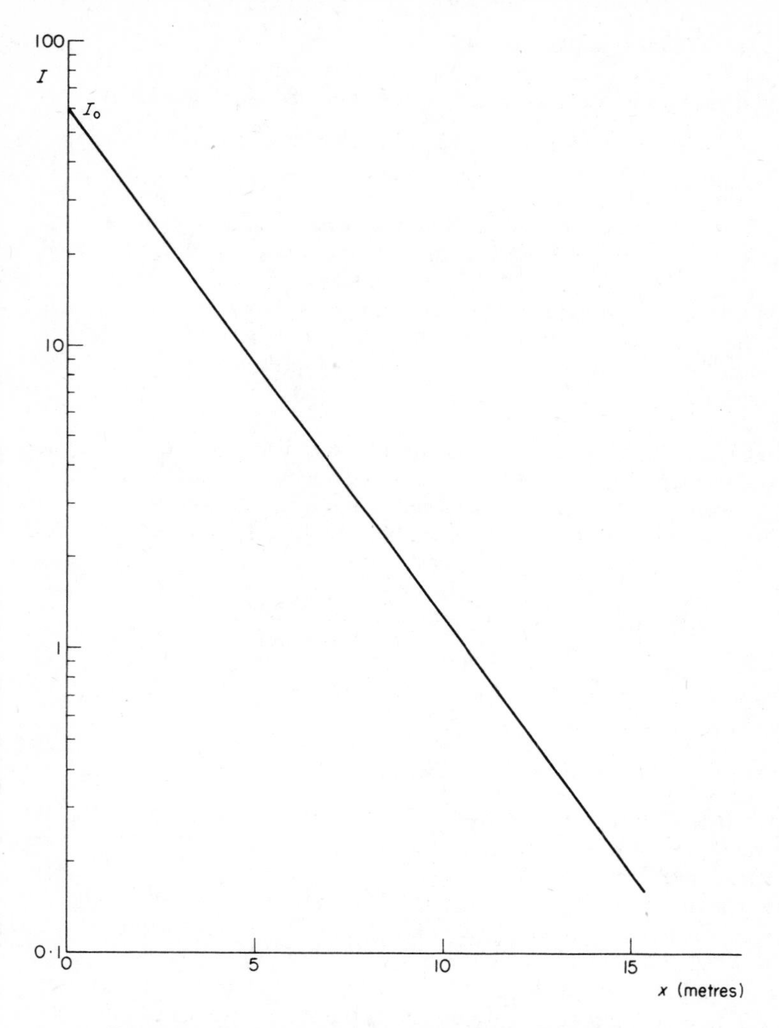

Figure 2.30 To illustrate the use of log-linear graph paper

(figure 2.29). Of course δI must be negative. If we assume that the energy absorbed between x and $x + \delta x$ is proportional to the distance travelled, and to the incident intensity I, then Lambert's law follows. We have in fact

$$\delta I = -\lambda I \delta x \tag{2.46}$$

where λ is a constant of proportionality. This gives

$$\frac{dI}{dx} = -\lambda I \tag{2.47}$$

which is equivalent to equation 2.45.

Because exponential attenuation is rather commonly found, there are some special techniques for dealing with it. We see from equation 2.45 that

$$\begin{aligned}\log_{10} I &= \log_{10} I_0 - \lambda x \log_{10} e \\ &= \log_{10} I_0 - 0.434\,\lambda x\end{aligned} \tag{2.48}$$

Logarithms to base 10 are used for practical reasons. Thus a graph of $\log_{10} I$ against x yields a straight line. It is convenient to draw a graph of this kind on *log-linear* graph paper, on which the horizontal axis has a linear scale, and the vertical axis has a logarithmic scale. An example is shown in figure 2.30. As shown there, we mark up the I axis in *decades* 0.1, 1, 10, 100, etc. Note that there is no point for $I = 0$, because $\log_{10} 0 = -\infty$. We then simply plot values directly onto the graph; on figure 2.30, for example, $I = 10$ at $x = 4.7$, $I = 1$ at $x = 10.7$ and so on.

The basic property of exponential attentuation is that in a given distance x_0, I always decreases by the same factor irrespective of its initial value. On figure 2.30, for example I decreases by a factor of 10 (from 60 to 6, or from 10 to 1, or whatever) in a distance of 6 m. This can be seen directly from equation 2.45; we have

$$\frac{I(x+x_0)}{I(x)} = \frac{I_0 \exp[-\lambda(x+x_0)]}{I_0 \exp(-\lambda x)} \tag{2.49}$$

The factor $I_0 \exp(-\lambda x)$ cancels, to give simply

$$\frac{I(x+x_0)}{I(x)} = \exp(-\lambda x_0) \tag{2.50}$$

Thus the factor by which I decreases depends only on the distance and not on the initial value $I(x)$. This is what we set out to prove. Because of this basic property, it is often convenient to use *decibels* (dB) to describe attentuation. If $I_2/I_1 = 0.1$, we say that the level of I_2 is -1 bel, or -10 dB, relative to I_1. Formally,

$$\text{Relative level in dB} = 10 \log_{10}(I_2/I_1) \tag{2.51}$$

On figure 2.30, I decreases by a factor of 10 in 6 m. Thus the attenuation is 1 bel, or 10 dB, in 6 m. We may therefore express the attenuation constant λ as

$$\lambda = 10/6 = 1.67\ \text{dB m}^{-1} \tag{2.52}$$

Note from equation 2.51 that the decibel is a measure of *relative* intensity I_2/I_1. This may seem to conflict with the common use of decibels as an apparently absolute measure of noise levels. For example, the noise of an aircraft taking off, heard by a nearby village, might be quoted, loosely, as 90 dB. The point is, that this is a measure relative to some base level, for example an assumed threshold for audibility. However, the subject of noise measurement is complicated, involving consideration of the frequency range covered by the noise and the sensitivity of the ear over that range, and we shall not deal with noise measurement here.

2.7 Tunnelling

It is an important feature of wave motion that the amplitude of a wave can never change abruptly in space; at most the amplitude changes over a few wavelengths. This is best shown by the phenomenon which we may call wave *tunnelling*.

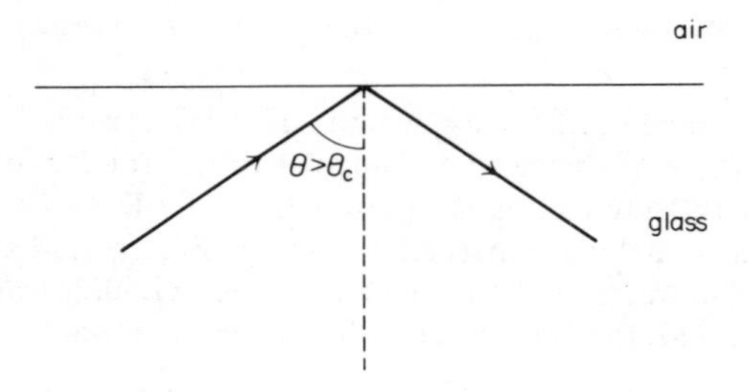

Figure 2.31 Total internal reflection

Consider light undergoing total internal reflection at a glass–air interface. It follows from Snell's law that if the angle of incidence θ in the glass is greater than a critical value θ_c, then the wave is reflected into the glass and there is no transmitted wave in the air (figure 2.31). However, the wave does not cease abruptly at the interface; its amplitude decays over a few wavelengths in the air. The wave motion in the air is called an *evanescent wave*. The existence of the evanescent wave can be shown if one brings up a second piece of glass to make an 'air sandwich' a few wavelengths thick, as shown in figure 2.32. The evanescent wave in the air is sufficiently intense to excite a weak transmitted wave T. We may say that the wave *tunnels* through the air.

The optical experiment we have just described is possible, but obviously difficult because of the short wavelength of the light. The tunnel wave can be demonstrated much more easily using 3 cm microwaves. Figure 2.33 shows a

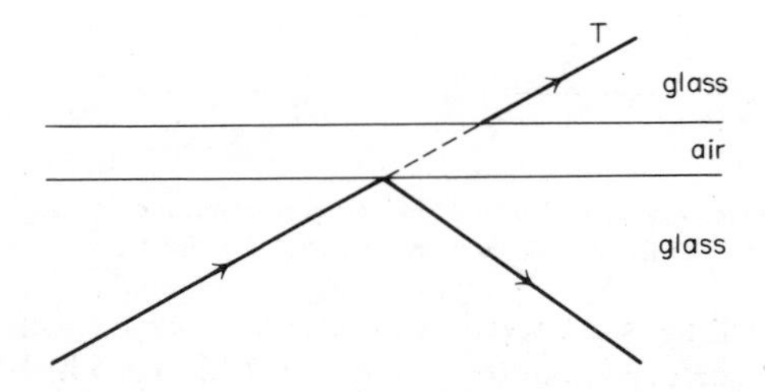

Figure 2.32 Air sandwich. A weak transmitted tunnel wave *T* appears if the air is only a few wavelengths thick

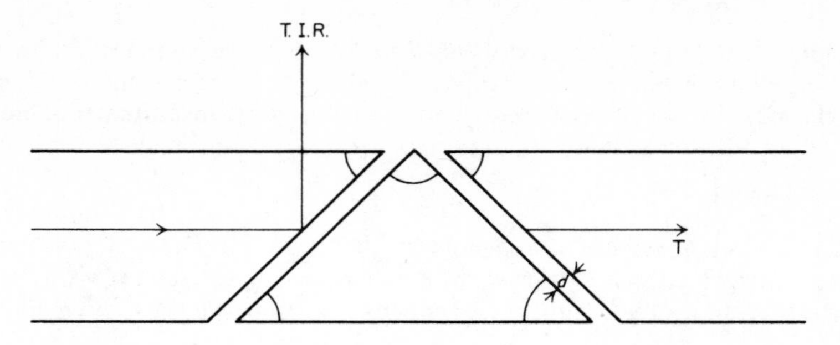

Figure 2.33 Microwave apparatus to demonstrate evanescent wave tunnelling.

suitable apparatus made of Perspex, which has a refractive index of 1.5 for microwaves. In the absence of the central triangular prism, the incident wave would be totally internally reflected along the path marked T.I.R. Once the prism is in place, however, a tunnel wave marked T appears. Figure 2.34 shows, as a function of the air gap d, the fraction of the incident power which is transmitted along T. As might be expected, the transmission decays exponentially as d is increased.

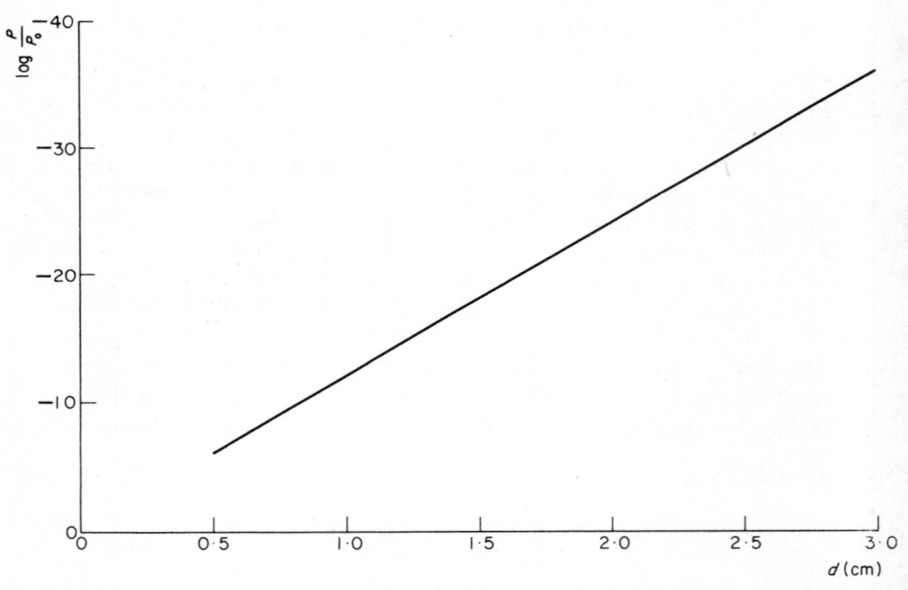

Figure 2.34 Transmitted power as a function of gap spacing d for apparatus of figure 2.33. Note logarithmic scale for power. (Courtesy of Mr L. Richards)

Whenever tunnelling is observed, we know that we are dealing with a wave of some kind. As we shall see in detail in chapter 4, things which in a simple way might be regarded as particles, such as electrons and protons, can behave as waves in some circumstances. In particular, they can tunnel through barriers, like the evanescent wave in figure 2.33. We may give two examples. First we deal with the electron 'tunnel junction' of figure 2.35. This is made by evaporating a metal strip M_1 typically 500 nm thick and 1 mm wide, allowing it to oxidise, and then evaporating a second strip M_2 across it. This produces a thin insulating oxide layer I

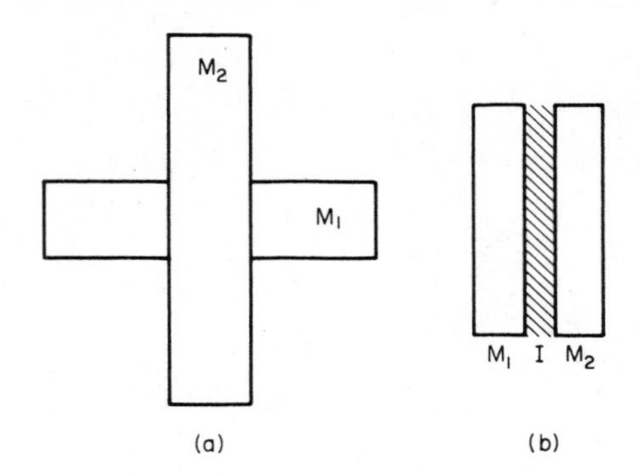

Figure 2.35 (a) Plan and (b) section of electron tunnel junction

between the two metals. The metal M_1 is most frequently aluminium, because the oxide is strong and stable. If I were sufficiently thick, there would be no current between M_1 and M_2. However, when I is between 1 and 10 nm thick, as is the case in practice, then some current flows by tunnelling of the waves associated with the electrons. Figure 2.36 shows the tunnel resistance R as a function of oxide thickness d; again R increases exponentially as d increases.

Secondly we turn to the α-particle decay of some heavy nuclei, which is an important example of tunnelling. An atomic nucleus consists of Z protons, of charge $+e$, and N neutrons, which are uncharged. Some heavy nuclei, for example

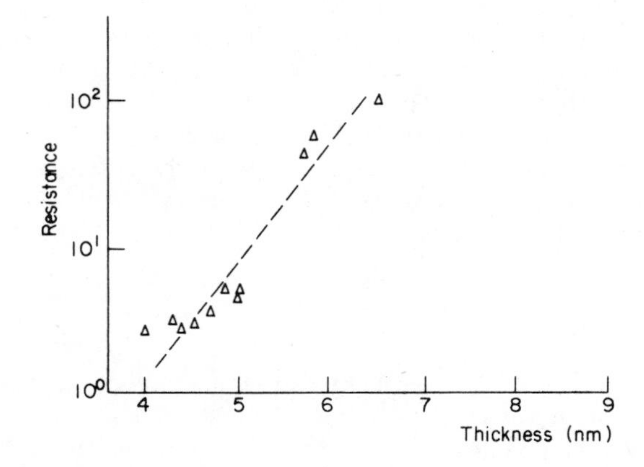

Figure 2.36 Tunnel resistance as a function of oxide thickness in Al–Oxide–Al junctions. The thickness is deduced from the capacitance of each junction. (After J. C. Fisher and I. Giaever *J. appl. Phys.*, **32** (1961), 172)

^{224}Ra, which contains 88 protons and 136 (= 224 – 88) neutrons, are *α unstable.* This means that from time to time a sample of the material emits an α-particle, or ^4He nucleus, which consists of two protons and two neutrons. It cannot be predicted at what time a given nucleus will decay, but the decay process has a characteristic *half life* $T_{1/2}$. After a time $T_{1/2}$, half the nuclei in a given sample will have

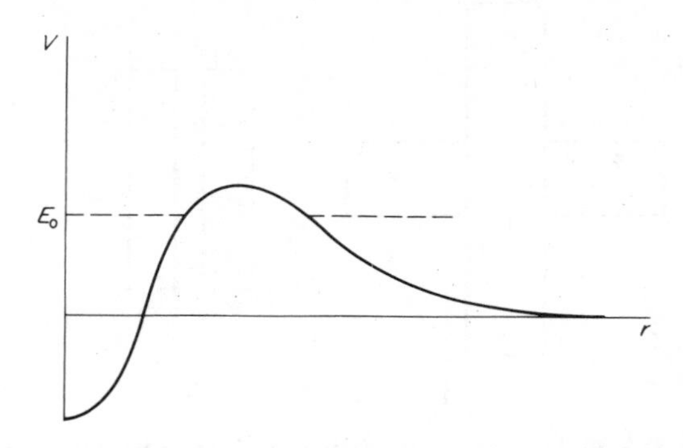

Figure 2.37 Potential energy of an α-particle as a function of distance from a nucleus

decayed. Half lives vary enormously; for example ^{232}Th has a half life of 14×10^{10} years, while ^{212}Po has a half life of 300 ns. It is instructive to sketch the potential energy V of an α-particle relative to the nucleus which is left after the decay. We take V to be zero at separation $r = \infty$, as shown in figure 2.37. The α-particle and the nucleus are both positively charged, so for large separations it takes work to move them closer together against the electrostatic, or Coulomb, force between them. Thus V increases as r decreases from ∞. However, as r approaches the characteristic range of the strong forces which bind the nucleus together, the force becomes an attractive one and the potential energy decreases. The complete picture, therefore, is of a *potential well* near the origin, together with a repulsive *Coulomb tail.* Now consider an α-particle which is bound in the well at an energy E_0 which is greater than zero, but less than the maximum in the potential energy curve. If the α-particle were nothing but a particle, it would stay in the well indefinitely with energy E_0. However, because the α-particle has a wave-like character as well, it can, and eventually does, tunnel through the hump in the potential energy curve. This is the process of α-particle decay, and we see that it depends crucially on the particle having a wave-like character.

2.8 Electromagnetic Spectrum

It was pointed out in section 2.3 that light is only one kind of electromagnetic radiation, that is, transverse radiation in which an electric and a magnetic field oscillate in the plane perpendicular to the direction of propagation. We have also seen that microwaves are electromagnetic radiation of a much longer wavelength than light. For convenience, we give here a brief description of the properties of electromagnetic radiation of various frequencies, as summarised in figure 2.38.

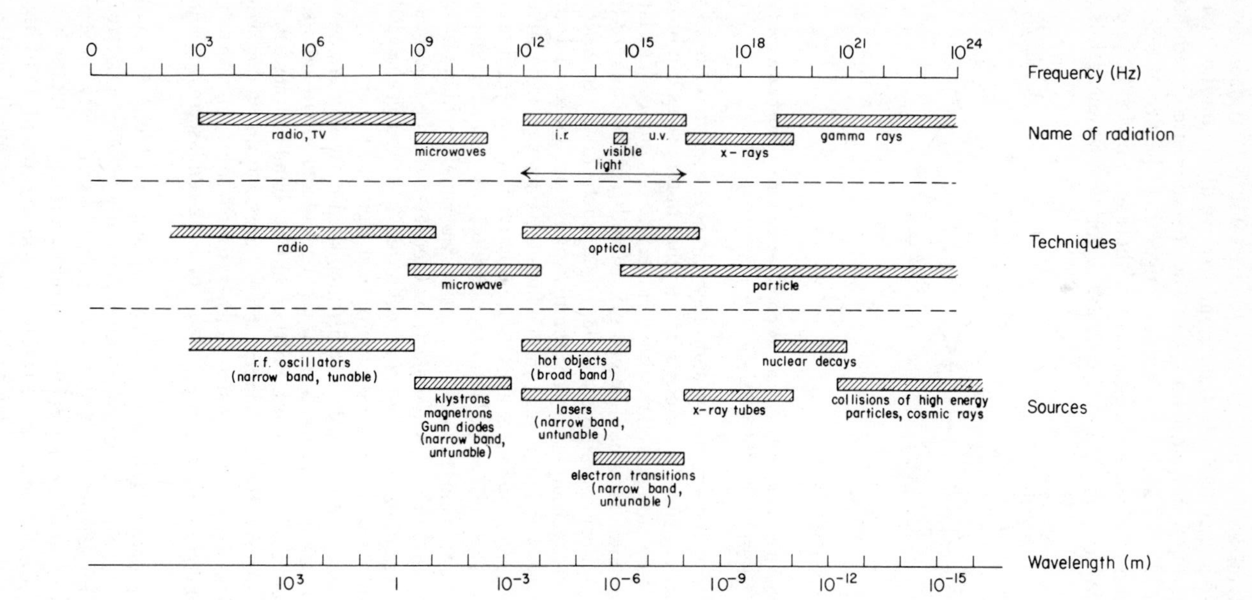

Figure 2.38 The electromagnetic spectrum

The first row of the figure simply gives the names for radiation of different frequencies; we may make the usual observation that visible light occupies a relatively narrow frequency band. In the second row we divide up the techniques used to handle radiation into four main types. The characteristic of radio technique is that the radiation wavelength is much longer than the size of the circuitry in the transmitter and receiver, and consequently there are no phase changes of the waves within the circuitry. In the microwave region, the wavelength is comparable with the size of the apparatus, and phase changes are important. Furthermore, the comparatively small wavelength means that microwaves can be, and usually are, transmitted along coaxial cable or metallic waveguides. The latter simply consist of rectangular metallic pipework, of width about equal to the wavelength. In the optical region, the wavelength is much shorter than the size of the apparatus, and the techniques are the familiar ones using lenses and mirrors. Finally, we distinguish a 'particle' region, in which wave packets travel like particles, or photons, and individual arrivals are observed in a detector such as a Geiger–Muller tube. The division between techniques is by no means sharp; for example, light can be transmitted along an optical waveguide, which consists of a thin layer (typically 500 nm) of one dielectric evaporated on the surface of another dielectric. Again, *photon counting* techniques are used as a specialised spectroscopic technique well down into the optical region.

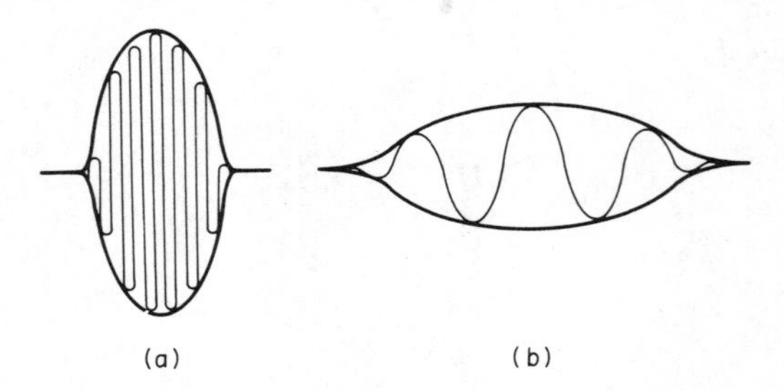

Figure 2.39 High-frequency radiation, (a), can be compressed into a short particle-like wave packet, whereas low-frequency radiation, (b), cannot

In the row on sources (figure 2.38) we distinguished between narrow-band sources, which produce more or less a single frequency, and broad-band sources, which produce a range of frequencies. Further, narrow-band sources can be tunable, which means that the frequencies can be varied, or untunable, with a fixed frequency. This division is not sharp either; for example, microwave sources are tunable by a few per cent, and there are some infra-red lasers which are tunable too.

Finally, we may take a preliminary look at the way in which a beam of high-frequency radiation, particularly gamma rays, generally behaves as a stream of particles. One characteristic of a particle is that it arrives in a very short time, or more exactly that it transfers energy and momentum to any detector in a time which is too short for the detector to resolve. Obviously, radiation of high frequency can be bunched into wave packets which have a much shorter duration than the corresponding wave packets for low-frequency radiation (figure 2.39).

The mere fact that gamma rays arrive as very short wave packets is, therefore, not surprising. However, the experimental evidence is that the wave packets always contain the same amount of energy, and it is this constancy in size that completely justifies the name *particle* for the wave packet. Chapter 4 will be devoted to the experimental evidence for the wave–particle duality we have just described and to some of the consequences of wave–particle duality.

Problems

1. What is a standing wave? Define the terms *node* and *antinode* in a standing wave, and give *three* examples from physical phenomena of standing waves, using different media in each case. [UE]

2. Polarisation—general case. Suppose the displacements are

$$u_x = a \cos \omega t + b \sin \omega t$$

$$u_y = c \cos \omega t + d \sin \omega t$$

so that we have arbitrary amplitudes and arbitrary relative phase. Show that, for any value of t

$$(au_y - cu_x)^2 + (du_x - bu_y)^2 = (ad - bc)^2$$

and hence that the polarisation vector (u_x, u_y) moves round an ellipse. In what directions are the principal axes of the ellipse?

3. Show that a quarter-wave plate converts the linearly polarised light

$$u_x = u_0 \cos \omega t$$

$$u_y = v_0 \cos \omega t$$

to elliptically polarised light. Why does a quarter-wave plate *always* convert circular polarisation to linear? [UE]

4. Show that a half-wave plate converts right-hand circularly polarised light to left-hand, and vice versa.

5. Explain precisely how a linear polariser and a quarter-wave plate can be used to make, (a) a left-hand circular polariser, (b) a right-hand circular polariser.

6. Show how a linear polariser and a quarter-wave plate can be used to construct a general elliptical polariser; that is, a device which will transform unpolarised light into elliptically polarised light of either handedness and with a variable ratio between the principal axes.

7. Polarisation by reflection. In general, when light falls on an air–glass interface, some light is reflected. However, when the E vector of the light is in the same plane as the direction of the incident wave and the normal to the surface, as an figure 2.40(a), and when the angle marked ϕ is 90°, the amplitude of the reflected wave is zero. On the other hand, with the polarisation shown in figure 2.40(b), there is always some reflection. What is the physical reason for this difference?

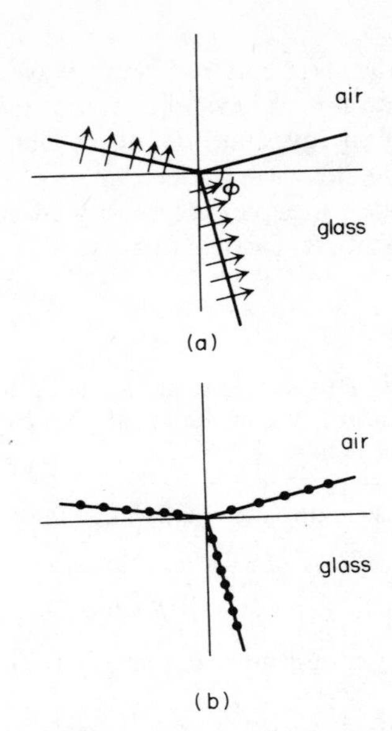

Figure 2.40 The two possible planes of polarisation for light falling on an interface

8. Show that when the angle ϕ in figure 2.40(a) is 90° the angle of incidence is $\theta_i = \theta_B$, where the *Brewster angle* θ_B satisfies

$$\tan \theta_B = n_2/n_1$$

9. The dispersion equation for waves on the surface of a liquid is

$$\omega^2 = gk + Ak^3$$

Derive expressions for v_g and v_p. Sketch ω, v_g and v_p as functions of k.

10. What are v_g and v_p as functions of k for a medium with dispersion

$$\omega = ck^2$$

Sketch ω, v_g and v_p as functions of k. [UE]

11. Show that the group velocity is given by

$$v_g = v_p - \lambda \frac{dv_p}{d\lambda}$$

12. Consider the superposition of two waves of slightly differing wavelengths λ_1 and $\lambda_2 > \lambda_1$. Using the result of the previous example, show that $v_g > v_p$ if the longer wavelength wave moves more slowly, and that $v_g < v_p$ if the longer wavelength wave moves more quickly. Interpret this result in terms of the movement of the nodes (points of opposite phase) in figure 1.10.

13. Consider a longitudinal wave travelling along the spring and mass system of figure 2.23. Let u_n be the displacement of the nth mass from its equilibrium position na. Show that the equation of motion for u_n is

$$m\frac{d^2u_n}{dt^2} = -C_1(u_n - u_{n-1}) + C_1(u_{n+1} - u_n)$$

$$= C_1(u_{n-1} - 2u_n + u_{n+1})$$

where C_1 is the elastic constant of every spring. Substitute the plane wave solution

$$u_n = A\ \exp(-i\omega t)\exp(ikna)$$

Show that this satisfies the equation of motion if

$$\omega^2 = \frac{2C_1}{m}(1 - \cos ka)$$

Defining $\rho = m/a$ as the density, and $C = aC_1$ as the elastic modulus show that this is the same as the dispersion equation in the text, equation 2.33 and figure 2.24.

14. Discuss the generation, propagation and detection of electromagnetic radiation of all frequencies.

15. Describe the possible states of polarisation of a transverse wave. Illustrate your discussion either mathematically, with formulae for the transverse displacements, or descriptively in terms of transverse waves on a stretched string. A transverse wave travelling in the z direction has transverse displacements at a particular value of z given by

$$u_x = A\ \cos(\omega t + \delta_1)$$
$$u_y = B\ \cos(\omega t + \delta_2)$$

What values must A, B, δ_1 and δ_2, have in order that the wave represents (a) general linear polarisation, (b) circular polarisation? Give sketches of the motion of the displacement vector (u_x, u_y) in each case. [UE]

16. Calculate the frequency separation between successive modes in a CO_2 laser 1 m long operating at a wavelength of 10.6 μm.

17. The following table gives the amplitude of swing A of a damped pendulum as a function of time t in seconds. For convenience, A is tabulated as A/A_0 where A_0 is the amplitude at $t = 0$.

t	0	2	4	6	8	10	12	14	16	18
A/A_0	1	0.7	0.44	0.31	0.19	0.14	0.10	0.065	0.042	0.03

By plotting the decay on log–linear graph paper, or otherwise, find the decay constant in decibels per second.

18. The dispersion equation for microwaves in a wave guide may be written

$$\omega^2 = c^2k^2 + \omega_c^2$$

where the *cut-off frequency* ω_c is a constant which characterises the waveguide. Show that the group and phase velocities satisfy

$$v_g v_p = c^2$$

Sketch the dispersion curve, and sketch v_g and v_p as functions of ω.

3

Diffraction

Any wave is *diffracted* by obstacles in its path. For example, figure 3.1 shows the optical diffraction pattern produced by a small hole in an opaque screen. Because of its wave character, the light is spread out after passing through the hole. Although diffraction is a general property of all waves, diffraction of light is the easiest to observe as well as the most important in practice, and we shall concentrate on optical examples.

To begin with, we distinguish between *Fresnel* diffraction (figure 3.2), in which source and detector are at a finite distance from the diffracting obstacle, and *Fraunhofer* diffraction (figure 3.3), in which source and detector are 'at infinity'; that is, in Fraunhofer diffraction we deal with an incident plane wave, and study the angular distribution of the emitted light. For a practical set up, one may use appropriate lenses, as sketched in figure 3.4. We shall restrict our attention to Fraunhofer diffraction; Fresnel diffraction may be treated similarly but the detailed calculations are more involved.

An essential condition for observing diffraction effects is that the radiation falling on the aperture must be coherent. As we saw in section 1.2, the practical requirement is normally that the light comes from a single source. As in section 1.2, we shall then be able to assume that the phase is the same across the incident plane wave front in Fraunhofer diffraction.

We shall calculate diffraction patterns with the aid of *Huygens' principle*, which states that we can find the diffraction pattern by treating each point of the aperture as an independent secondary source, radiating waves with a phase

Figure 3.1 Pinhole diffraction pattern. The central bright region is masked out to avoid fogging of the photographic plate. (Reproduced with permission from M. Cagnet, *et al.* *Atlas of Optical Phenomena*, Springer, Berlin (1962))

equal to that of the incident wave. The proof, which is due to Kirchhoff, is not particularly difficult; it uses primarily the fact that the governing equation for light is linear, so that one can use superposition. However, the proof is beyond the mathematical level of this book, so we shall simply take Huygens' principle for granted. It is worth noting that Huygens' principle is consistent with the

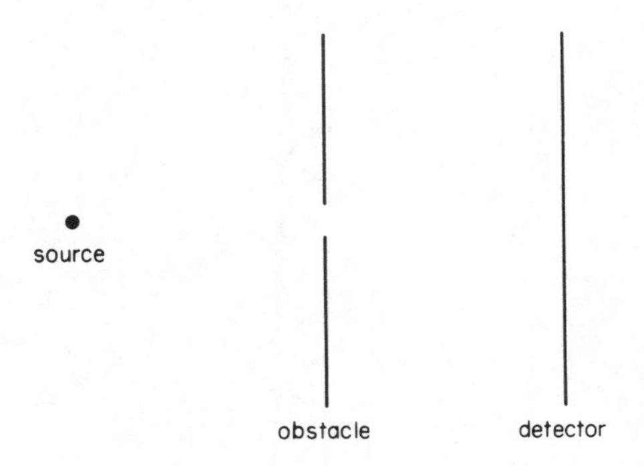

Figure 3.2 Fresnel diffraction

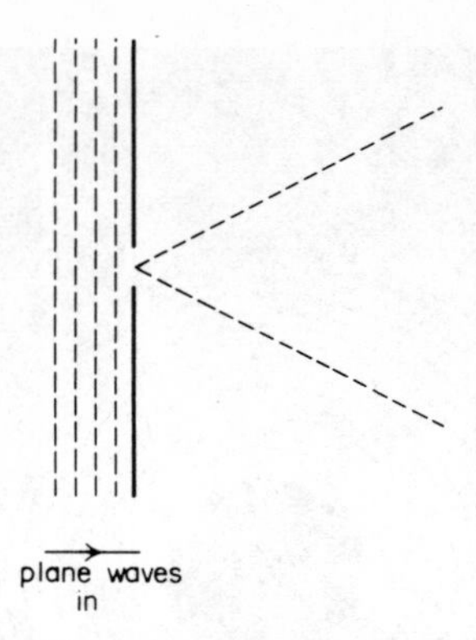

Figure 3.3 Fraunhofer diffraction

ordinary laws of geometrical optics. For example, if we have a plane wave propagating, the secondary sources lie on a plane. At a later time, the points of equal phase, that is, the points of the wave front, still lie on a plane, displaced an

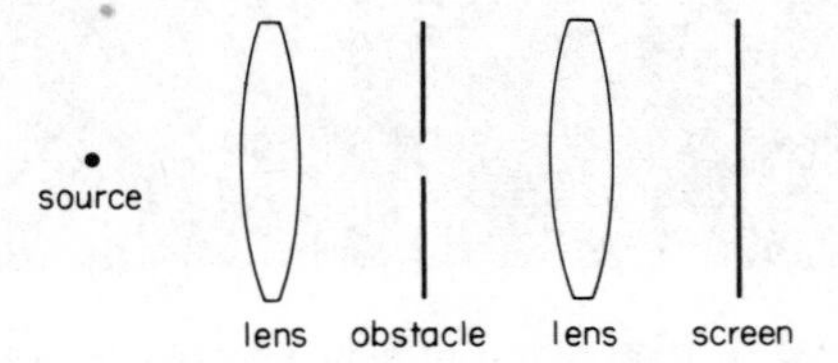

Figure 3.4 Practical arrangement for Fraunhofer diffraction

appropriate distance (figure 3.5). Thus the plane wave propagates in a straight line, as we know from geometrical optics. In a similar way, the equal angle law for reflection, and Snell's law of refraction, can be derived from Huygens' principle.

Figure 3.5 Straight-line propagation as a consequence of Huygens' principle

3.1 Single-Slit Diffraction Pattern

We may now calculate specific diffraction patterns. First we find the Fraunhofer diffraction pattern produced by a single slit in an opaque screen. As shown in figure 3.6, we take the z axis as the direction of travel of the incident plane wave. The slit of width d runs from $x = -d/2$ to $x = d/2$, and we assume that the slit is infinite in the y direction, so that figure 3.6 is the view of the slit in the x–z plane. We wish to calculate the intensity of the diffracted wave travelling in the direction which makes an angle θ with the z axis. At a distance L from the slit, the amplitude of the wave emitted by a secondary source of length $\mathrm{d}x$ centred at 0, the mid-point of the slit, may be written

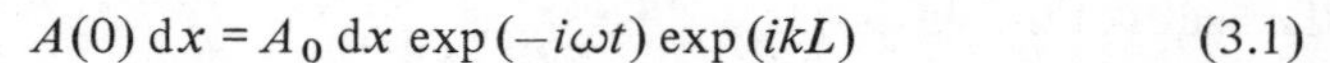

$$A(0)\,\mathrm{d}x = A_0\,\mathrm{d}x\,\exp(-i\omega t)\exp(ikL) \qquad (3.1)$$

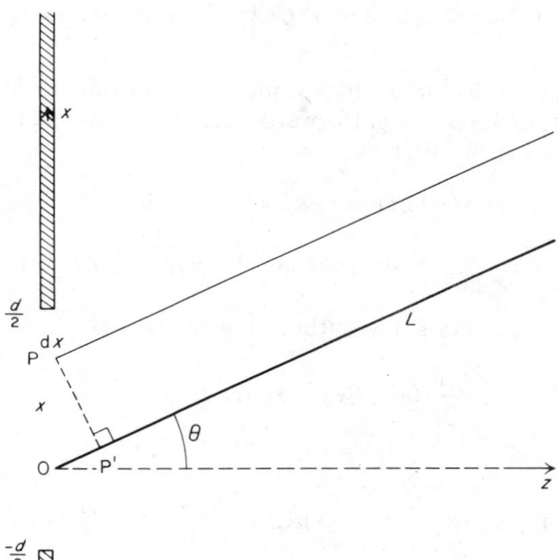

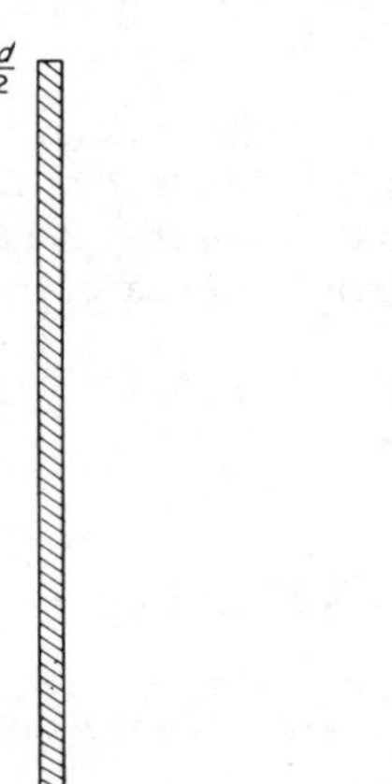

Figure 3.6 Notation for calculation of single-slit diffraction pattern

Here we are using the convention, explained in appendix 1, that we write the wave as the real part of a complex quantity, which means that the complex number A_0 contains both the amplitude and the phase of the wave emitted at 0. The secondary wave emitted from an element dx at the point P, a distance x from 0, travels a path which is shorter by the distance OP′ than that travelled by the wave from 0. Since the length OP′ is $x \sin\theta$ the wave from P differs in phase from the wave from 0 by a factor $\exp(-ikx \sin\theta)$, so we may write the wave from P as

$$A(x)\,dx = A_0\,dx \exp(-i\omega t) \exp(ikL) \exp(-ikx \sin\theta) \tag{3.2}$$

The total amplitude in the direction θ is found by integrating equation 3.2 over the width of the slit

$$A = A_0 \exp(-i\omega t) \exp(ikL) \int_{-d/2}^{d/2} \exp(-ikx \sin\theta)\,dx \tag{3.3}$$

where we have taken the factors independent of x outside the integral. The integral in equation 3.3 is straightforward, since the integrand is simply an exponential function. We find

$$A = A_0 \exp(-i\omega t) \exp(ikL) \times \frac{1}{ik \sin\theta} [\exp(\tfrac{1}{2}ikd \sin\theta) - \exp(-\tfrac{1}{2}ikd \sin\theta)] \tag{3.4}$$

We may simplify equation 3.4 with the aid of the identity

$$\frac{1}{2i} [\exp(ix) - \exp(-ix)] = \sin x \tag{3.5}$$

We find

$$A = A_0 \exp(-i\omega t) \exp(ikL) \frac{2}{k \sin\theta} \sin(\tfrac{1}{2} kd \sin\theta) \tag{3.6}$$

Now as we explain in appendix 1, the power crossing unit area in the direction θ at time t is proportional to the square of the amplitude, that is, the square of the real part of equation 3.6. The constant of proportionality depends on the medium through which the medium is travelling. We therefore have, for the power at time t

$$P(t) \propto B^2 \cos^2(\omega t - kL - \delta_0) \tag{3.7}$$

where we have put

$$B = A_{00} \frac{2}{k \sin\theta} \sin(\tfrac{1}{2} kd \sin\theta) \tag{3.8}$$

and expressed the complex amplitude A_0 in terms of a real amplitude and phase angle

$$A_0 = A_{00} \exp(i\delta_0) \tag{3.9}$$

The intensity I of the light travelling in the direction θ, that is the average flux of power, measured in watt m^{-2}, is given by the time average of the instantaneous power $P(t)$. Using equation 3.7, we therefore find

$$I \propto \tfrac{1}{2}B^2 \tag{3.10}$$

or with the aid of equation 3.8

$$I \propto A_{00}{}^2 \frac{2}{k^2 \sin^2 \theta} \sin^2 (\tfrac{1}{2} kd \sin \theta) \tag{3.11}$$

Equation 3.11 is the result we wanted for the diffraction pattern, that is, the intensity as a function of angle. In order to make it more intelligible, it is convenient to define a new variable

$$p = \tfrac{1}{2} kd \sin \theta \tag{3.12}$$

or in terms of the wavelength

$$p = \frac{\pi d}{\lambda} \sin \theta \tag{3.13}$$

We can write I in terms of p as

$$I = I_0 \frac{\sin^2 p}{p^2} \tag{3.14}$$

where we have absorbed the constant of proportionality which appears in equation 3.11 into I_0

$$I_0 \propto \tfrac{1}{2} A_{00}{}^2 d^2 \tag{3.15}$$

Since the limit as $p \to 0$ of $\sin p/p$ is 1, we see from equations 3.12 and 3.14 that I_0 is the intensity in the forward direction, $\theta = 0$.

It is a straightforward matter to sketch the intensity I as a function of the variable p. We have already seen that I_0 is the value at $p = 0$. Because of the presence of the sine function, I is zero at the points $p = n\pi$ for $n \neq 0$, and because of the factor p^2 in the denominator, the overall intensity falls off as p increases. The exact form of I is shown in figure 3.7(a). There is a *principal maximum* in the forward direction, and *subsidiary maxima* on either side. The width of the central maximum, from $-\pi$ to $+\pi$, is twice the width of each subsidiary maximum. The angular distance θ_m from the centre to the first minimum, that is half the width of the principal maximum, is given by $p = \pi$, that is

$$\sin \theta_m = \lambda / d \tag{3.16}$$

The angle θ_m gives a measure of the angular spread of the diffraction pattern.

Equation 3.16 is very important, because it shows the way in which the angular spread of the diffraction pattern scales with the wavelength λ and slit width d. We see first that as d increases θ_m decreases. The wider the slit, the less spread out is the diffraction pattern. This relation between expansion in distance and decrease of angular spread is common to all diffraction effects, and we shall see shortly that it is a consequence of the $\delta k \delta x$ uncertainty product which we mentioned in

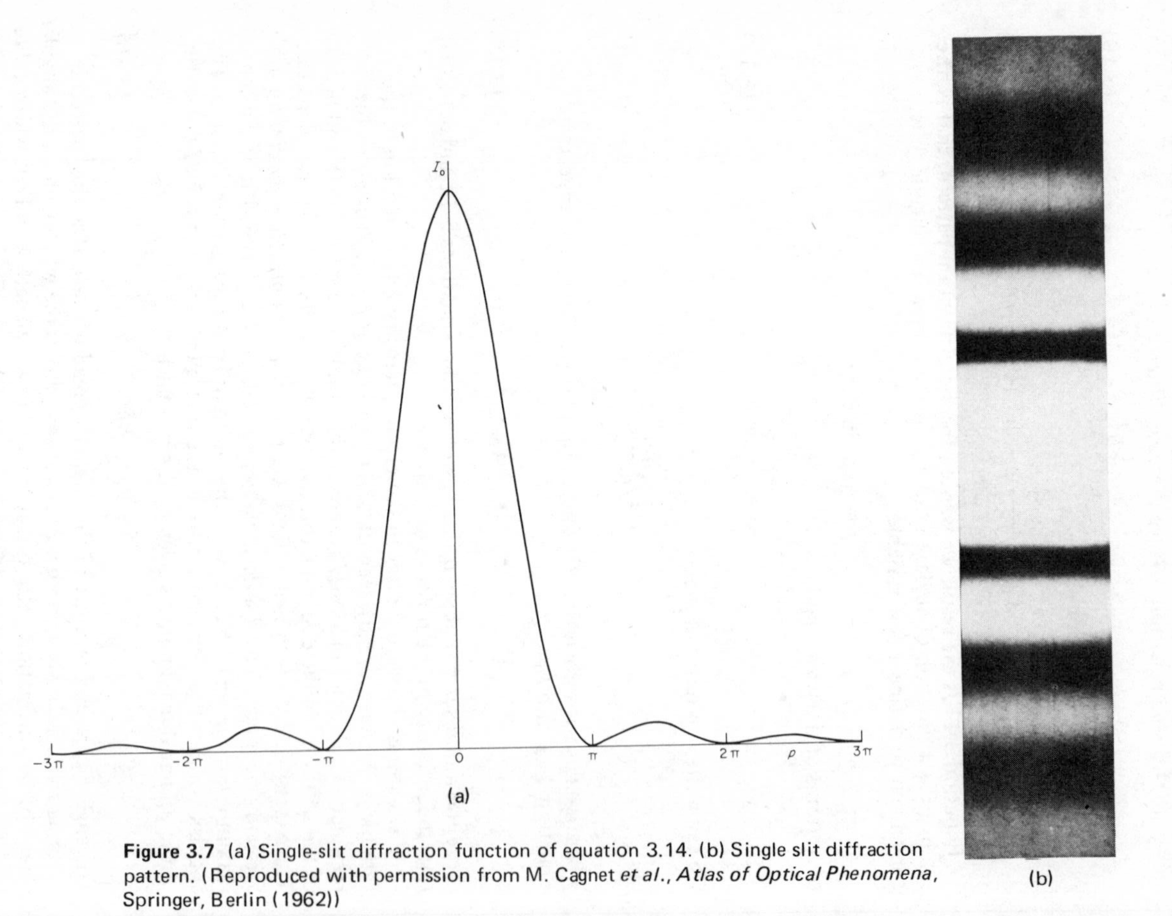

Figure 3.7 (a) Single-slit diffraction function of equation 3.14. (b) Single slit diffraction pattern. (Reproduced with permission from M. Cagnet *et al.*, *Atlas of Optical Phenomena*, Springer, Berlin (1962))

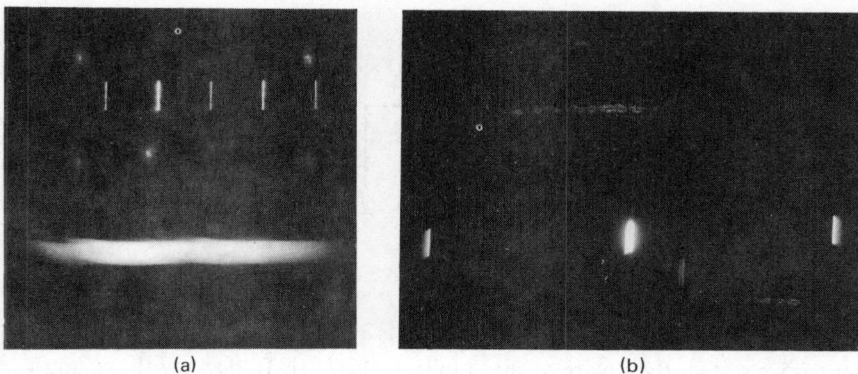

(a) (b)

Figure 3.8 Single-slit diffraction patterns for two different slit widths; slit for pattern (a) set narrower than slit for (b)

section 2.1. An example of the decrease of θ_m as d increases is shown in figure 3.8, where we compare single-slit diffraction patterns for two different slit widths. For $d \gg \lambda$, which of course is usual for diffraction using optical light, the overall angular spread of the diffraction pattern is small, and we may use the small-angle approximation $\sin \theta_m = \theta_m$ to write

$$\theta_m = \lambda/d \qquad \text{for small angles} \tag{3.17}$$

We note that when d is so great that diffraction effects are negligible, equation 3.17 gives $\theta_m \approx 0$; there is no sideways spread, and the beam travels in the forward direction. This is simply rectilinear propagation, which as we have already mentioned is the simplest consequence of Huygens' principle.

The wavelength dependence in equation 3.16 can also be verified experimentally. The fact that the angle is governed by the ratio λ/d follows from a simple dimensional argument. The diffraction problem is defined by the two lengths λ and d, and the angular spread can only depend on the dimensionless combination λ/d.

3.2 Relation of Single-Slit Pattern to Square-Wave Frequency Spectrum

The single-slit diffraction pattern of figures 3.7 and 3.8 bears a striking resemblance to the frequency spectrum of a square pulse, shown in figure 1.17(b). In fact, since the frequency spectrum is the amplitude $|\tilde{u}(\omega)|$, and the diffraction pattern is the intensity, which is the square of the amplitude, the diffraction pattern is the square of the frequency spectrum. Our aim in this section is to see why this equivalence holds. Suppose we express the single slit of figure 3.6 in terms of an *aperture function* $u(x)$, which will be defined so that $u(x) = 1$ where the screen is transparent, and $u(x) = 0$ where the screen is opaque. The aperture function of the single slit is therefore a *square pulse* (figure 3.9). We can show that the amplitude of the diffracted wave in the direction θ is given by the frequency spectrum $|\tilde{u}(k_x)|$, where $|\tilde{u}(k_x)|$ is related to $u(x)$ just as $|\tilde{u}(\omega)|$ is related to $u(t)$, and k_x is

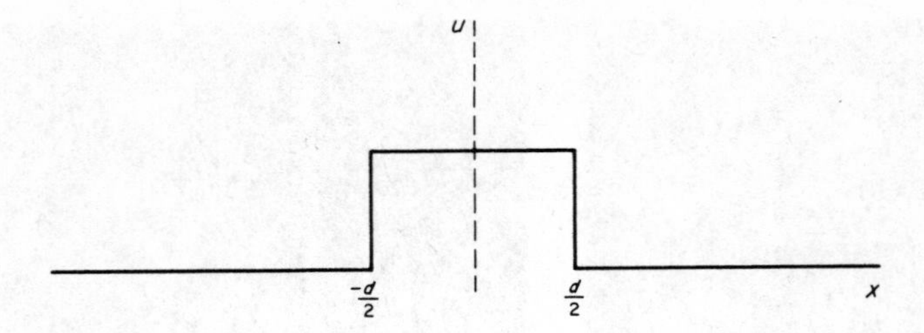

Figure 3.9 Aperture function for a single slit

identical to the variable p defined in equations 3.12 and 3.13. The intensity is then, as we stated, the square of the amplitude $|\hat{u}(k_x)|$. To begin with, we must look a little further into the meaning of the wave numbers k_x and k_z.

In chapter 2, we dealt with waves travelling solely in one direction, which we took to define either the x or the z axis. Now that we are dealing with diffraction, we are interested in waves travelling in a variety of directions. We continue to use

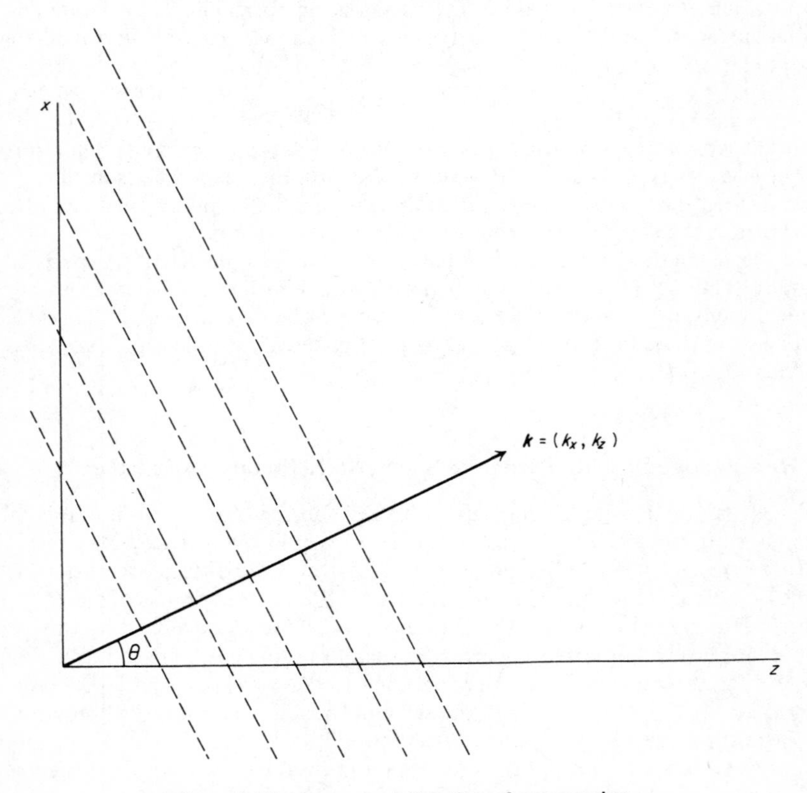

Figure 3.10 Vector k as direction of propagation

the co-ordinate system of figure 3.6, and we write the diffracted wave in some direction as

$$u(x, z, t) = u_0 \exp(-i\omega t) \exp(ik_x x + ik_z z) \tag{3.18}$$

The key result is that the vector $\boldsymbol{k} = (k_x, k_z)$ gives the direction of travel of the wave. The wavefronts, the successive crests or troughs of the wave, are the planes along which the phase of the wave is constant, and the direction of travel is perpendicular to the wavefronts (figure 3.10). We can see from equation 3.18 that the planes of equal phase are given by

$$k_x x + k_z z = constant \tag{3.19}$$

and the direction at right angles to these planes is indeed that specified by the vector $k = (k_x, k_z)$. We can also relate k_x and k_z to the wavelength λ. We write

$$k_x = k \sin\theta \tag{3.20}$$

$$k_z = k \cos\theta \tag{3.21}$$

where θ is the angle which the vector $\boldsymbol{k}$ makes with the z axis, and k is the magnitude of $\boldsymbol{k}$. If the point (x, z) moves a distance λ in the direction specified by the angle θ, then the phase $k_x x + k_z z$ increases by 2π. This gives

$$k_x \lambda \sin\theta + k_z \lambda \cos\theta = 2\pi \tag{3.22}$$

or with the values given by equations 3.20 and 3.21

$$k\lambda(\sin^2\theta + \cos^2\theta) = 2\pi \tag{3.23}$$

that is

$$k = 2\pi/\lambda \tag{3.24}$$

That is, the magnitude of the vector $\boldsymbol{k}$ is related to the wavelength in the usual way. Figure 3.11 summarises the results we have just obtained.

The identification of $\boldsymbol{k}$ with the direction in which the wave is travelling holds good also in three dimensions. We shall see a further significance in the result in chapter 4, when we discuss the motion of the particle which according to quantum mechanics is associated with a wave packet. We shall find that the momentum $\boldsymbol{p}$ is proportional to the wave number $\boldsymbol{k}$, and of course $\boldsymbol{p}$ defines the direction of travel.

We can now return to the single-slit diffraction pattern. Because of the analogy between t and ω on the one hand, and x and k_x on the other, which we pointed out in section 2.1, the aperture function of figure 3.9 corresponds to a distribution of k_x values $|\tilde{u}(k_x)|$ which is given by the function of figure 1.17(b). Using equation

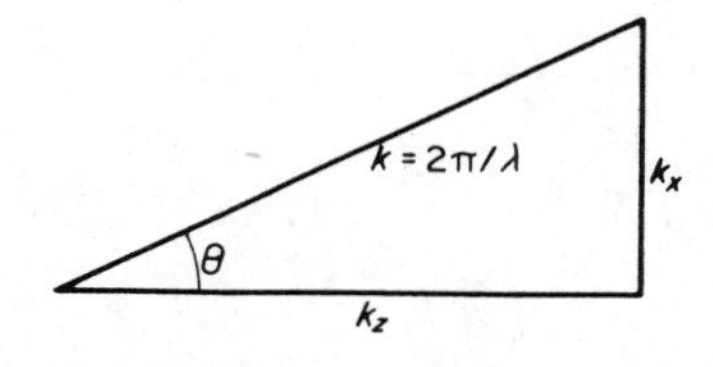

Figure 3.11 Relation between components and magnitude of $\boldsymbol{k}$

3.20 to introduce the angle θ, we find that the amplitude at angle θ is given by $\tilde{u}(k \sin \theta)$, and hence that the intensity is

$$I \propto |\tilde{u}(k \sin \theta)|^2 \tag{3.25}$$

This is the result which we stated at the beginning of this section.

It is easy to see, from our present point of view, how equations 3.16 and 3.17 for the width of the diffraction pattern arise. We saw in section 2.1, equation 2.3, that the width δk_x of the function $|\tilde{u}(k_x)|$ is given by the uncertainty relation $\delta k_x \delta x \approx 2\pi$. In the present case, δx is the width d of the slit, so we have

$$\delta k_x \approx 2\pi/d \tag{3.26}$$

With the use of equations 3.20 and 3.24 we may convert this into an angular width

$$\delta k_x = 2\pi\delta(\sin \theta)/\lambda \tag{3.27}$$

Combining equations 3.26 and 3.27, we have

$$\delta(\sin \theta) \approx \lambda/d \tag{3.28}$$

which is the same as equation 3.16, although expressed in slightly different notation. In particular, the decrease of θ_m as d increases, shown in figure 3.8, is a straightforward consequence of the uncertainty relation $\delta k_x \delta x \approx 2\pi$.

3.3 Applications of Single-Slit Pattern

The results of the previous section imply that whenever we pass a beam through a finite aperture, a diffraction pattern is formed. All optical instruments have a finite aperture, so diffraction sets a limit on their resolving power. We may consider two examples–the resolving power of an optical telescope, and the production of a diffraction-limited beam by a source of finite dimensions.

Suppose we look at two distant stars through a telescope. Each star appears not as a point, but rather as the diffraction pattern formed by the end of the telescope. The question is, how far apart in angular distance do the stars have to be before the telescope *resolves* them, that is, separates one diffraction image from the other? The problem can be slightly simplified by regarding the end of the tele-

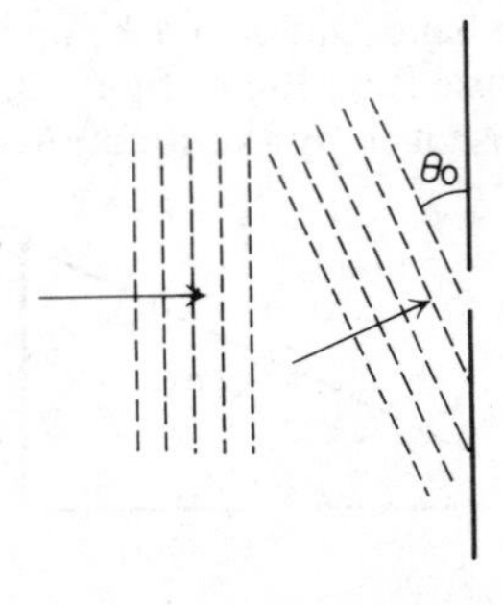

Figure 3.12 Two plane waves incident on a slit

scope as the single slit of the previous sections, rather than a circular aperture. Now suppose that as well as a plane wave at normal incidence, there is a plane wave incident at angle θ_0 (figure 3.12). It is easy to see that in the small-angle approximation, which requires only $d \gg \lambda$, the second wave produces a diffraction

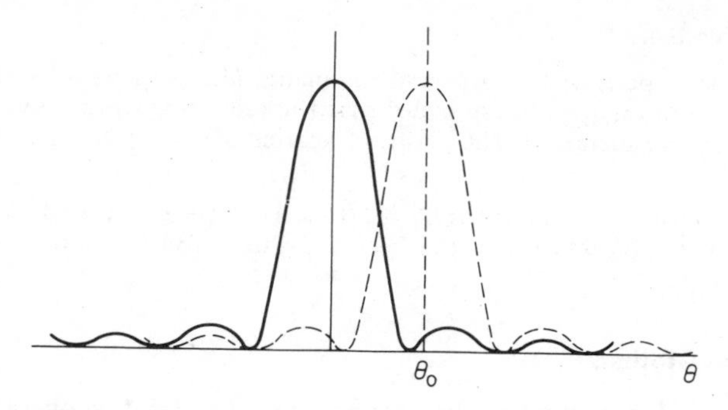

Figure 3.13 Diffraction patterns produced by the waves of figure 3.12.

pattern centred at $\theta = \theta_0$ (figure 3.13). Whether or not the diffraction patterns can be seen to be distinct depends to some extent on the aptitude of the observer. It is convenient to have a simple objective rule for assessing whether the diffraction patterns are distinct; such a rule is *Rayleigh's criterion*, which states that the beams are resolved if the principal maximum of the oblique beam falls at or beyond the first minimum of the normally incident beam. Using equation 3.17, we see that the beams are resolved according to Rayleigh's criterion if

$$\theta_0 > \lambda/d \tag{3.29}$$

As might be expected, the greater the aperture d, the better the resolution.

For our second example, we may consider the form of beam emitted by a source of finite diameter, such as the circular end face of a laser. Since the light emerges only from the finite aperture formed by the source, the beam produced is actually the diffraction pattern of the source, and is called a *diffraction-limited beam* (figure 3.14). With a circular source, such as the end face of a laser, the beam

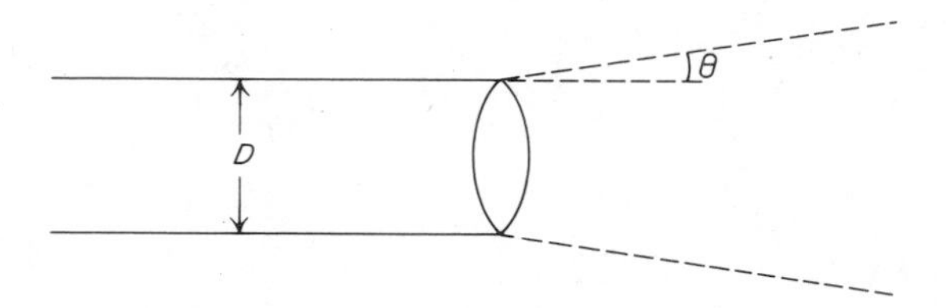

Figure 3.14 Diffraction-limited beam

has the form of the pinhole diffraction pattern shown in figure 3.1. We can see from the general arguments of the previous two sections that the angular spread of the beam must be of order λ/D, where D is the source diameter. An exact calculation for a circular source shows that the angular width from the centre to the first dark fringe is $1.22\ \lambda/D$. The spreading with distance of the beam produced by a laser of visible frequency can readily be seen in practice.

Worked example 3.1

An infra-red telescope is used to observe the planet Mars at a wavelength of 50 μm. Estimate the minimum aperture in order that the telescope should resolve structure of scale 1 km on the surface of the plant. (Distance of Mars = 8×10^7 km.)

Answer

The angular scale of the structure is $\delta\theta = 1/(8 \times 10^7)$ (linear scale divided by distance) So we require $\lambda/d < \delta\theta$ where d is the aperture. Putting in the numbers, $d > \lambda/\delta\theta$, or $d > 4$ km.

3.4 Diffraction Gratings

We now turn to the important practical example of a diffracting obstacle consisting of N slits, each of width d and a distance D apart. To be precise, D is taken to be the distance between the mid-points of adjacent slits, as shown in figure 3.15.

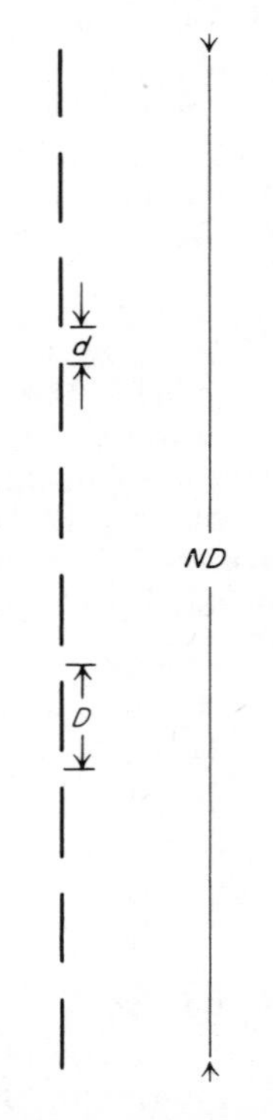

Figure 3.15 A diffraction grating

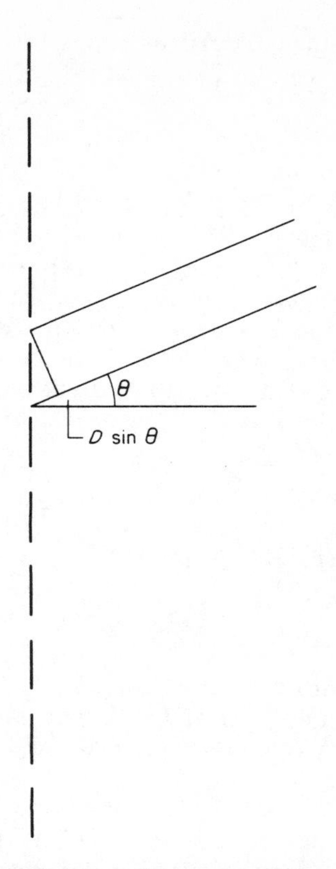

Figure 3.16 Path-length difference in diffraction grating

What we have is a *diffraction grating*, which we shall see is very important in spectroscopy.

We can see qualitatively what to expect for the diffraction pattern. The grating involves three lengths, $\delta x \approx d, D, ND$ respectively, so there are three significant ranges of k_x, with spreads $\delta k_x \approx 2\pi/d, 2\pi/D, 2\pi/ND$. As can be seen from figure 3.11, the corresponding angles are given by $\sin\theta \approx \lambda\delta k_x/2\pi$, that is $\sin\theta \approx \lambda/d, \lambda/D, \lambda/ND$. We may expect these to be the angles that feature in the answer.

We now give a more precise argument. The difference in optical path length between light from neighbouring slits is $D \sin\theta$ (figure 3.16). Therefore when $D \sin\theta = m\lambda$, or

$$\sin\theta_m = m\lambda/D \qquad (m = \text{integer}) \tag{3.30}$$

all the contributions from all the slits are in phase, so we expect a maximum in intensity. If the number of slits N is very large, as is generally the case in practice, we may expect the peak at θ_m to be very sharp. The reason is that if θ moves a little from the reinforcement condition $D \sin\theta = m\lambda$ the phase changes by 2π from one end of the array to the other, and we see cancellation, not reinforcement.

Therefore the width of the peak may be estimated as follows. Let $\theta = \theta_m + \delta\theta$. The shift in phase from one slit to the next is

$$D \sin(\theta_m + \delta\theta)\frac{2\pi}{\lambda} = (D \sin\theta_m + D\delta\theta \cos\theta_m)\frac{2\pi}{\lambda} \tag{3.31}$$

where we have used the formula for $\sin(A + B)$ from appendix 1, and made the small-angle approximations $\cos\delta\theta \approx 1$, $\sin\delta\theta \approx \delta\theta$. Because θ_m satisfies the reinforcement condition of equation 3.30, the phase change given by the first term in equation 3.31 is $2m\pi$ which is unimportant. The phase change across the whole array is N times the second term, that is, $ND\delta\theta \cos\theta_m 2\pi/\lambda$. When this total phase change is 2π, the waves from all the slits combine destructively, so that the intensity is zero. Thus the half width of the peak at θ_m, from the centre to the zero on either side, is

$$\delta\theta_m = \frac{\lambda}{ND\cos\theta_m} \tag{3.32}$$

For fairly small values of θ_m, this is simply

$$\delta\theta_m = \frac{\lambda}{ND} \tag{3.33}$$

A more detailed analysis shows, as is plausible, that there are small subsidiary maxima associated with the peak at θ_m. These are found for values of $\delta\theta$ slightly larger than that given by equation 3.33. However, the subsidiary maxima are generally unimportant.

Up to now the variation in the phase of the wave across each slit has been ignored. This variation has the effect that the intensity of emission at angle θ is not the same as in the forward direction, but is given by the single-slit diffraction pattern of equation 3.14 and figure 3.7. The first minimum in this pattern is at

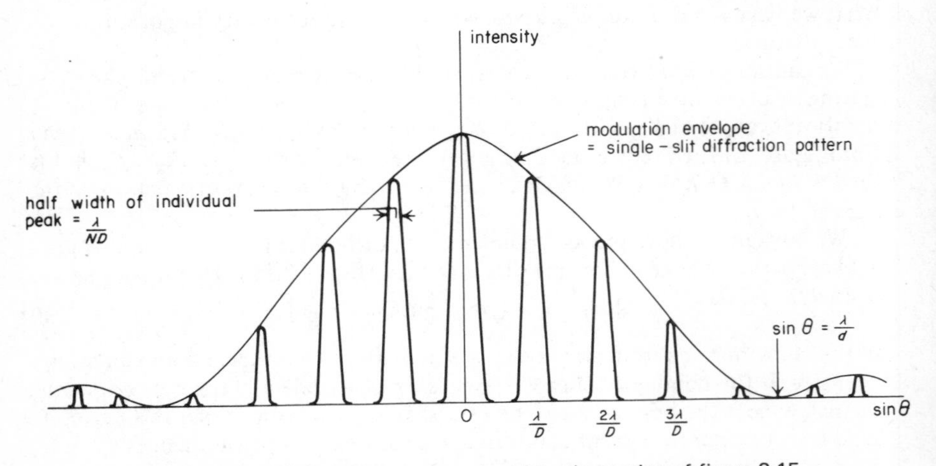

Figure 3.17 Diffraction pattern produced by the grating of figure 3.15

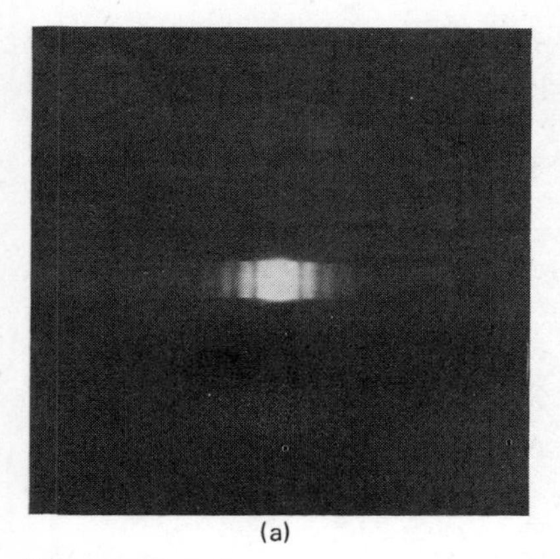

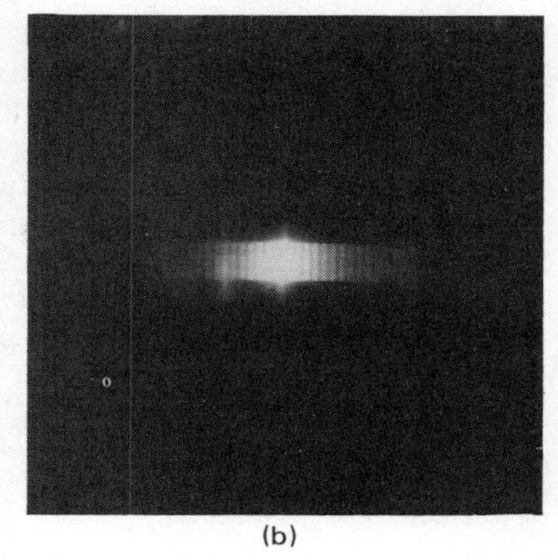

Figure 3.18 Diffraction patterns from gratings of two different ruling spacings: (a) 2000; (b) 7500 lines per inch

$\sin\theta = \lambda/d$. The angular spacing between the sharp maxima is λ/D, which is much smaller, so the single-slit diffraction pattern acts as a slowly varying modulation.

Our results for the grating diffraction pattern are summarised in figure 3.17. As anticipated, the pattern is governed by the three values $\sin\theta \approx \lambda/d$, λ/D, λ/ND. These give respectively the scale of the slow modulation, the position of the sharp diffraction peaks, and the width of each peak.

The dependence on D of the spacing of the diffracted lines is easily demonstrated; figure 3.18 shows diffraction patterns from two gratings of different line spacing. As always in diffraction, the larger the line spacing D, the smaller the angular spread of the diffraction pattern.

We observed that the single-slit diffraction pattern resembled the frequency spectrum of an isolated square wave (figure 1.17(b). Similarly it can be seen that there is a strong resemblance between the diffraction pattern of figure 3.17 and the frequency spectra of trains of square pulses shown in figures 1.15(b) and 1.16. As we pointed out for the single-slit pattern, the diffraction is an intensity, and is therefore the square of the frequency spectrum, which is an amplitude. The reason for the similarity can be seen by looking at the aperture function for the diffraction grating. Just as the single slit was represented by the square aperture function of figure 3.9, so the diffraction grating can be represented by the repeated square aperture $u(x)$ of figure 3.19. This is the spatial equivalent of a sequence of square pulses, and as shown in section 3.2, the diffraction pattern is simply given by $|\tilde{u}(k\sin\theta)|^2$, where $\tilde{u}(k\sin\theta)$ is the frequency spectrum as defined in section 1.4.

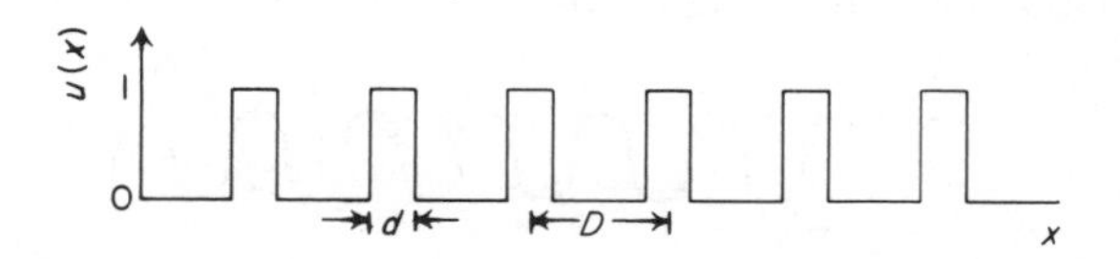

Figure 3.19 Aperture function for grating of figure 3.15

A diffraction grating is used in spectroscopy to resolve lines of different frequency. For example, sodium emits two closely spaced narrow lines in the yellow part of the visible spectrum, one at 589.6 nm, one at 589.0 nm. These are the lines which give the characteristic colour of a sodium street lamp, or of salt in a flame. If the yellow light from sodium falls on a diffraction grating, two diffraction patterns are produced, one spaced by λ_1/D, and the other spaced by λ_2/D. We may ask whether or not these are resolvable and again we use Rayleigh's criterion. The spectra are *resolvable in order m* if the peak at $m\lambda_2/D$ falls outside the half width λ_1/ND of the peak at $m\lambda_1/D$. This requires

$$\frac{m\lambda_2}{D} - \frac{m\lambda_1}{D} > \frac{\lambda_1}{ND} \tag{3.34}$$

or putting

$$\delta\lambda = \lambda_2 - \lambda_1 \tag{3.35}$$

the condition is

$$\delta\lambda > \frac{\lambda_1}{mN} \tag{3.36}$$

It is customary to define the *chromatic resolving power* as $\lambda/\delta\lambda$, that is

$$\frac{\lambda}{\delta\lambda} = mN \tag{3.37}$$

It is worth noting that the resolving power depends on the order m and the total number of lines N, but is independent of the ruling distance D.

We have now dealt with the properties of the simple transmission diffraction grating of figure 3.15, to which corresponds the aperture function of figure 3.19. It is described as a transmission grating because the diffraction pattern is observed on the further side from the incident beam. In practice, transmission gratings often consist of periodic rulings on transparent plastic. Such a grating does not have any opaque sections; but it is characterised by some periodic aperture function, as in figure 3.20. The features of the diffraction pattern produced by this grating which depend on the ruling distance D and the total width ND of the grating are the same as in the diffraction pattern of the simple grating we treated in detail. That is, the spacing λ/D between the peaks and their half width λ/ND are the same as in figure 3.17. However, the modulation envelope, which depends on the angular pattern produced by the individual slit, may be expected to be different.

In infrared spectroscopy in particular it is common to use a reflection grating, which consists of a reflecting surface with periodic rulings; the reflected light forms a diffraction pattern similar to that of figure 3.17. It is sometimes advan-

Figure 3.20 Aperture function for a general transmission grating

tageous to use a blazed reflection grating. As shown in figure 3.21, the surface of a blazed grating is angled so that most of the incident light is reflected into one particular order, say m, of the diffraction pattern. Since the resolving power increases with m, according to equation 3.37, the blazed grating has a greater

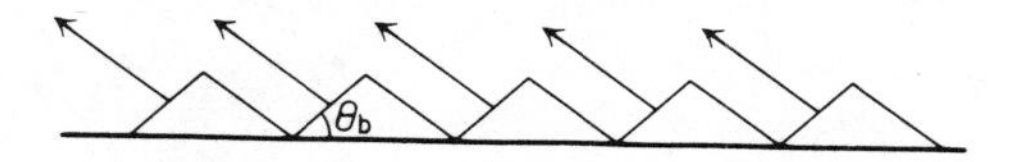

Figure 3.21 Blazed reflection grating

resolving power than an ordinary grating working on a low order line. However, it can be seen from equation 3.30 that the direction in space of the mth order line, and therefore the blaze angle of the grating, depends on the wavelength λ. The blazed grating therefore operates, with its enhanced resolving power, only for a certain band of wavelengths which is determined by the blaze angle and the desired order m. This link between an increase of sensitivity and a decrease of bandwidth is very common in all kinds of instruments.

Worked example 3.2

In section 2.2 we quoted the width of a line in Nd-doped glass as 4 THz. Assuming the line is centred at 600 nm, what is the minimum number of lines on a diffraction grating to resolve the line in first order?

Answer

The resolving power is given in equation 3.37 in terms of $\lambda/\delta\lambda$. We are given the linewidth in frequency as $\delta f = 4$ THz. Since $c = f\lambda$ we have

$$0 = f\delta\lambda + \lambda\delta f$$

or

$$\frac{\delta\lambda}{\lambda} = -\frac{\delta f}{f}$$

The frequency of a line at $\lambda = 600$ nm is

$$f = \frac{c}{\lambda} = \frac{3 \times 10^8}{6 \times 10^{-7}} = 5 \times 10^{14}$$

We therefore have

$$\frac{f}{\delta f} = \frac{5 \times 10^{14}}{4 \times 10^{12}} = 125$$

In order that the line be just resolved in first order, equation 3.37 gives $N = |\lambda/\delta\lambda| = |f/\delta f|$ (modulus sign because δf is the difference between the high and low frequency parts of the line). Thus

$$N = 125$$

Worked example 3.3

A blazed grating is to be constructed with spacing D to concentrate light into order m at wavelength λ. Show that the blaze angle, marked θ_B in the sketch below, is given by

$$\theta_B = \tfrac{1}{2} \sin^{-1}(m\lambda/D)$$

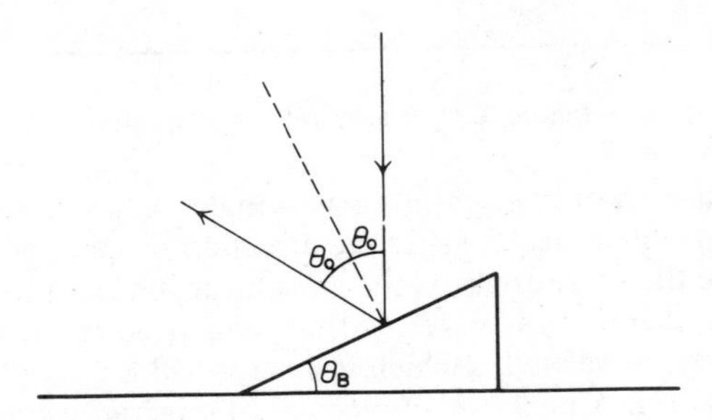

Answer

The angular direction of the mth maximum is given by equation 3.30: $\sin \theta_m = m\lambda/D$. Reflection off the blaze implies $\theta_0 = \tfrac{1}{2}\theta_m$, and from the geometry $\theta_0 = \theta_B$. Thus $\theta_B = \tfrac{1}{2}\theta_m$, which is the result required.

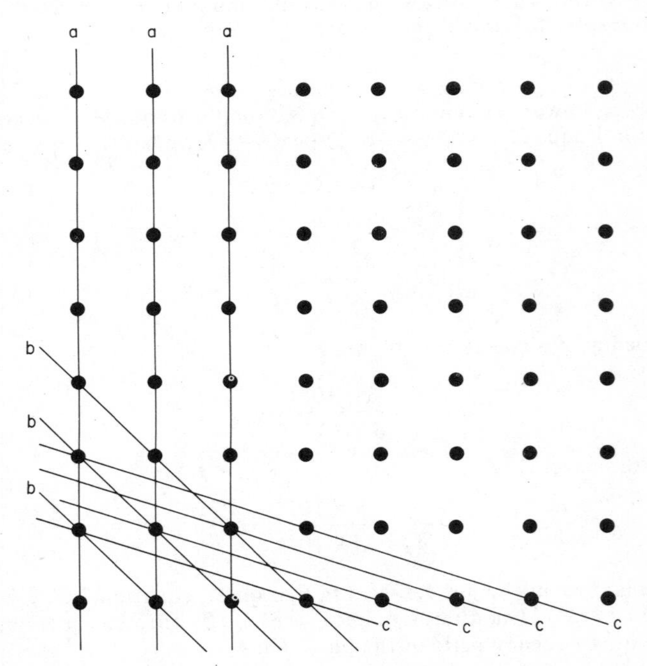

Figure 3.22 Three sets a, b, c of equally spaced lines in a two-dimensional square lattice

3.5 Bragg Scattering

Up to now we have described optical diffraction gratings in which the ruling distance D is comparable with, or somewhat greater than, the optical wavelength λ. Diffraction effects can also be observed with X-rays on crystals, since the spacing of successive planes of atoms in a crystal is typically 1 nm or less, which is the same order as X-ray wavelengths. Since the diffracting obstacle is now three dimensional, the description of the diffraction is somewhat more elaborate.

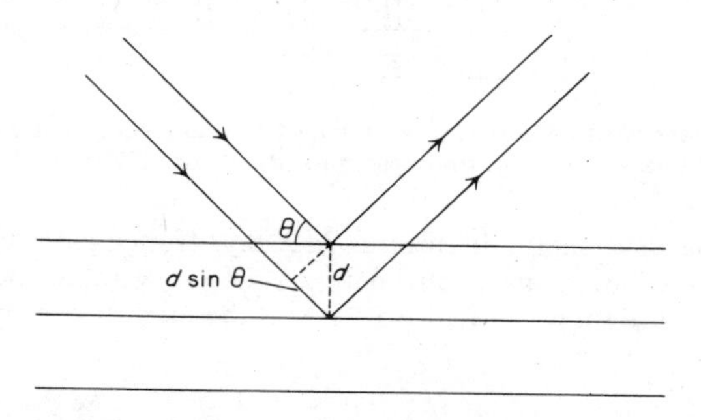

Figure 3.23 Path-length difference in Bragg scattering

The process of most interest is the reflection of light from a set of equally spaced planes in the crystal which are inclined at some angle θ to the incident light. As illustrated in figure 3.22 for a two-dimensional lattice of points, there are many possible ways of finding equally spaced sets of planes in a crystal lattice. Each set of such planes forms a kind of diffraction grating, and we shall now see how constructive interference can occur between waves scattered at the different planes of a set.

Consider the wavefronts reflected from two successive planes of a set with spacing d, as shown in figure 3.23. The path difference between the waves is $2d \sin \theta$ where θ is the angle marked, and the waves are in phase if this is a whole number of wavelengths. If the waves from these two planes are in phase, then so are all the waves from the planes of the equally spaced set, so the condition for constructive interference from all the planes is

$$2d \sin \theta = n\lambda \tag{3.38}$$

This is known as *Bragg's law.*

Bragg's law implies that for X-rays to be reflected, d, λ and θ must be related by equation 3.38. Monochromatic X-rays, that is, X-rays of a single wavelength, pass straight through the crystal except for angles of incidence θ for which a set of planes can be found with a lattice spacing d satisfying equation 3.38. The reflection of monochromatic X-rays off one set of planes therefore, has the appearance of the diffraction pattern of figure 3.24, with reflection only for particular angles θ.

Bragg reflection can be used in a number of different ways. The first application, developed by Bragg himself, involved using a crystal of known structure

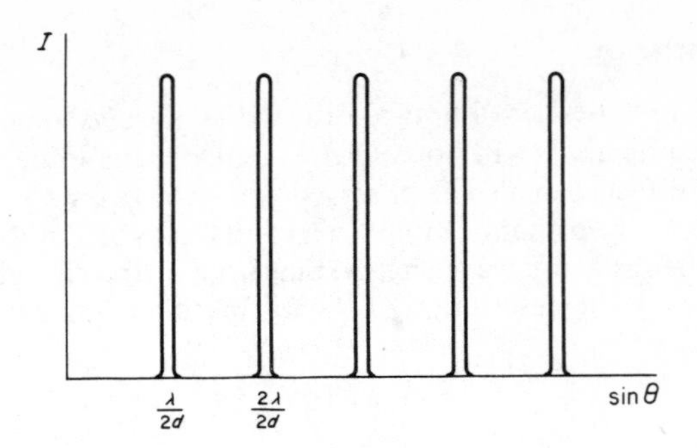

Figure 3.24 Intensity of reflection (I) versus sin θ for monochromatic X-rays and a set of lattice planes with spacing d. θ is the angle marked in figure 3.23

to measure the wavelength distribution in X-rays from a particular source; that is, Bragg constructed an X-ray spectrometer. A single crystal of rock salt, NaCl, has the cubic structure shown in figure 3.25, and the interatomic spacing a can be

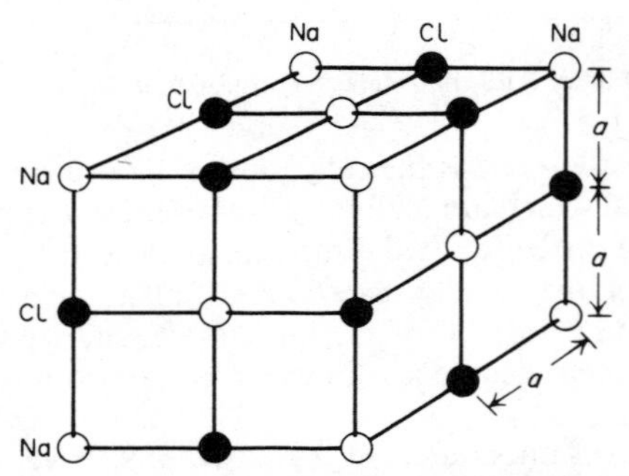

Figure 3.25 Crystal structure of rock salt, NaCl. (Reproduced, with permission, from T. A. Littlefield and N. Thorley, *Atomic and Nuclear Physics,* 2nd edn, van Nostrand, London (1968))

calculated from the density and the molecular weight. The spacings d of the various sets of equally spaced planes can be found from simple geometrical arguments. The spectrometer is arranged, as shown in figure 3.26, so that the

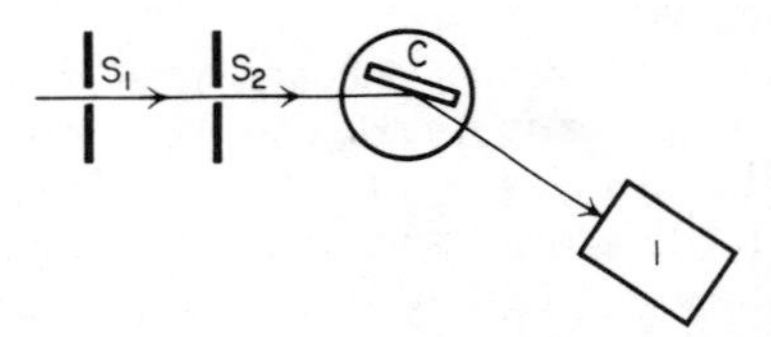

Figure 3.26 Bragg's X-ray spectrometer. S_1 and S_2 are collimator slits, C is the crystal of rock salt, and I is an ionisation chamber which detects the reflected X-rays. (Reproduced, with permission, from T. A. Littlefield and N. Thorley, *Atomic and Nuclear Physics,* 2nd edn, van Nostrand, London (1968))

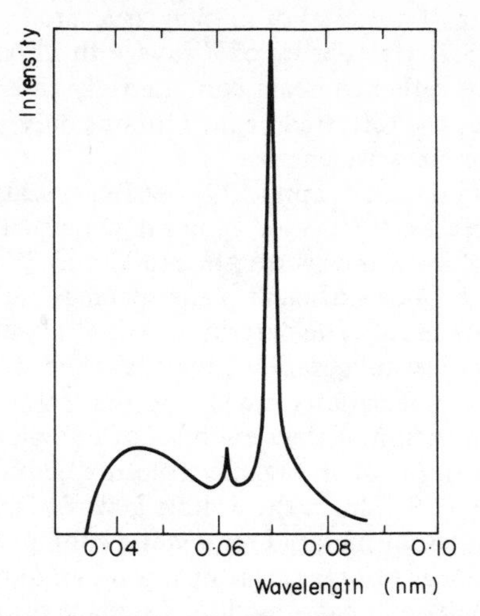

Figure 3.27 Intensity versus wavelength for X-rays from a Mo target bombarded by 30 keV electrons. (Reproduced, with permission, from C. Kittel, *Introduction to Solid State Physics*, 3rd edn, Wiley, New York (1966))

crystal can be rotated relative to the incident X-ray beam, and the ionisation chamber is rotated through twice the angle of the crystal, in order to detect any reflected X-rays. Equation 3.38 gives the conversion from the angles θ at which reflection occurs to the corresponding wavelengths λ.

X-rays are generally produced by bombardment of a metal target with high energy electrons, and it is found that the X-ray spectrum consists of sharp lines which are characteristic of the element, superimposed upon a continuous background. An example is shown in figure 3.27. A second application of Bragg

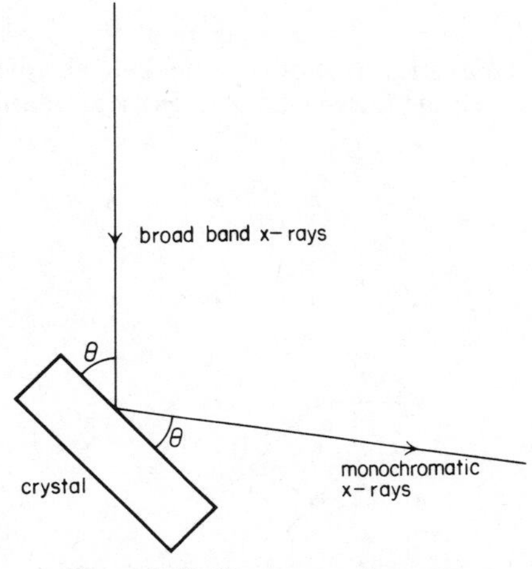

Figure 3.28 X-ray monochromator

reflection involves the use of a crystal of known structure to construct an X-ray monochromator (figure 3.28). If a beam of X-rays with a continuous spectrum falls on a crystal, then the reflected beam contains only those wavelengths satisfying equation 3.38. Thus the reflected beam contains only one wavelength, or more generally a few discrete wavelengths.

The third, and most widespread, application of Bragg scattering is the determination of crystal structures. If a crystal of unknown structure is exposed to monochromatic X-rays of known wavelength, equation 3.38 relates plane spacings d to angles of reflection θ. Once sufficient plane spacings are known, then the crystal structure can be deduced. The determination of crystal structures in this manner is known as X-ray crystallography. Nowadays the structures of all inorganic crystals of importance have been determined, and the technique is sufficiently well developed for the determination of the structures of complex organic molecules. An early, and famous, example of an organic structure deduced partly from X-ray results is shown in figure 3.29. This is the double helix of DNA (deoxyribonucleic acid), which is the fundamental material for genetic coding. The two helices (spirals) consist of identical sequences of organic groups, going in opposite directions as shown by the arrows. Each helix has a random sequence of bases (G, C, A, T in the figure) attached to it, and the helices are held together by hydrogen bonds between the bases. The structure was proposed by J. D. Watson and F. H. C. Crick, and an entertaining account of its discovery, together with further details of the structure, is given in the book *The Double Helix* by J. D. Watson (Weidenfeld and Nicholson, 1968).

Worked example 3.4

Suppose the two-dimensional square lattice, figure 3.22, has spacing a between nearest neighbour points. Show that the interplane spacing d for the planes marked b in figure 3.22 is

$$d = a/\sqrt{2}$$

Answer

Two successive lines of the series are shown in the sketch below. We see that $d = a\cos\theta$ where θ is the angle which the lines make with the y axis. Since $\theta = \pi/4$ we have $d = a/\sqrt{2}$.

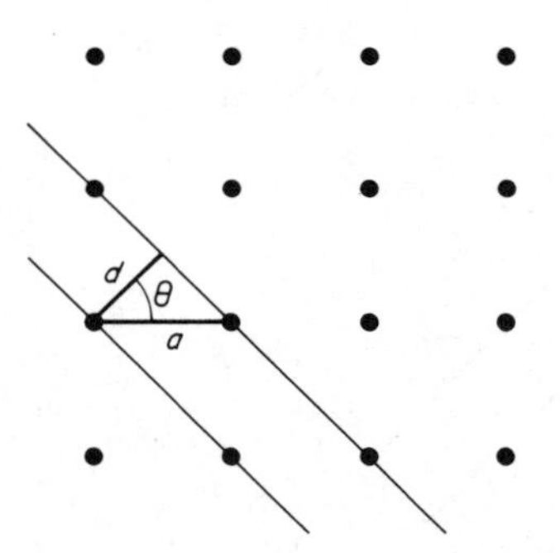

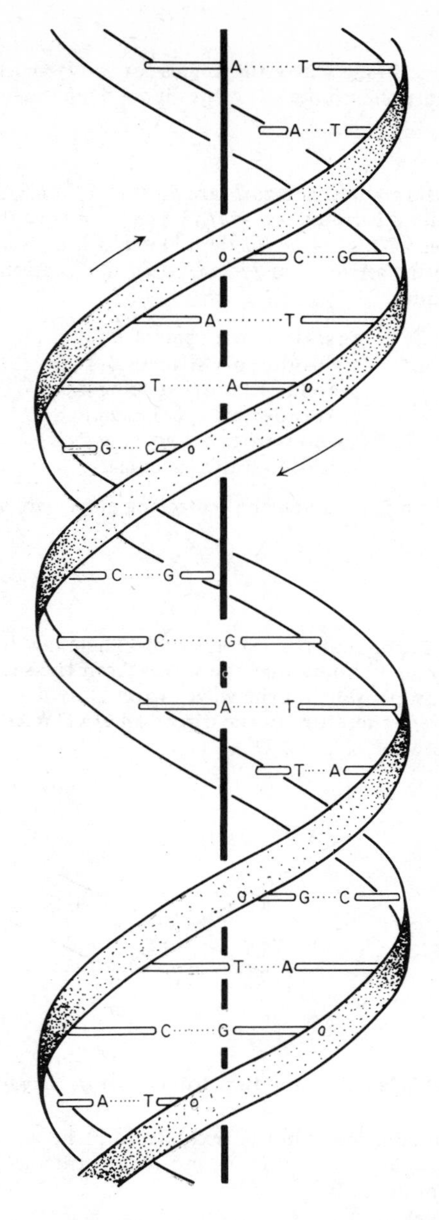

Figure 3.29 The double helix of DNA. (Reproduced, with permission, from J. D. Watson, *The Double Helix*, Weidenfield and Nicholson, London (1968))

Worked example 3.5

Find the six smallest Bragg scattering angles for the two-dimensional lattice of figure 3.22. What are the numerical values if $a = 1$ nm, $\lambda = 0.1$ nm?

Answer

The two largest interplane spacings d, are $d_0 = a, d_1 = a/\sqrt{2}$ (using previous example). The general formula is $d = a/(l^2 + m^2)^{1/2}$ (problem 3.15) so the sequence continues with $d_2 = a/\sqrt{(2^2 + 1^2)} = a/\sqrt{5}$, $d_3 = a/\sqrt{(2^2 + 2^2)} = a/2\sqrt{2}$. The general scattering angle is $\sin\theta = n\lambda/2d$ from equation 3.38, so we have in increasing magnitude

$\sin\theta = \lambda/2a$	(first-order off spacing d_0)
$\lambda\sqrt{2}/2a$	(first-order off spacing d_1)
λ/a	(second-order off spacing d_0)
$\lambda\sqrt{5}/2a$	(first-order off spacing d_2)
$\lambda\sqrt{2}/a$	(second-order off spacing d_1 *and* first order off d_3)
$3\lambda/2a$	(third-order off spacing d_0)

With the values given for λ and a the scattering angles are $2^\circ 52'$, $4^\circ 04'$, $5^\circ 44'$, $6^\circ 25'$, $8^\circ 08'$, $8^\circ 37'$.

Problems

1. It was stated in the text that Snell's equal angle law for reflection follows from Huygens' principle. Show that the waves from the secondary source at O′ in the sketch differ in phase from the waves from O by d sin θ/λ, and that the reflected wave is therefore in the direction OA. (What is the path-length difference between OA and O′A′?)

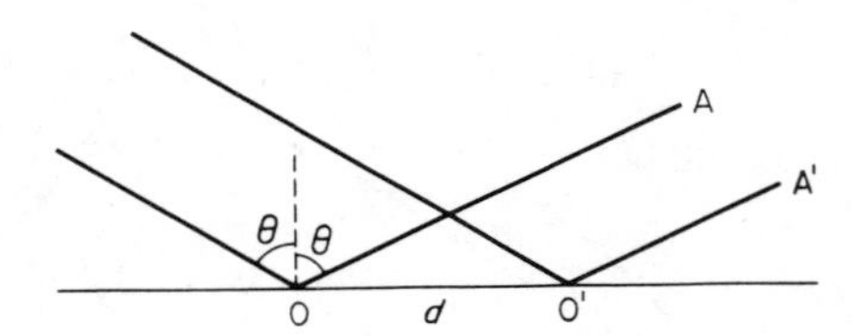

2. Show that Snell's law of refraction follows from Huygens' principle.

3. Prove the statement made in the text, that a plane wave falling on a single slit at a small angle of incidence θ_0 produces a diffraction pattern with its principal maximum at θ_0.

4. A diffraction-limited laser beam of diameter 10 m is pointed at the moon. What is the diameter of the area illuminated on the moon? The moon is 385 000 km away, and the wavelength of the laser is 632.8 nm.

5. Estimate the minimum aperture of nautical telescope required in order to read semaphore signals from a distance of 10 km.

6. Estimate the angular resolving power for a telescope of aperture 0.5 m operating at visible wavelengths. What scale of distance does this correspond to (a) on the surface of the moon (distance 385 000 km) (b) on the surface of the sun (distance 1.5×10^8 km), (c) at the distance of the nearest star (4.2 light years) (d) at the distance of the nearest galaxy (2.2×10^6 light years).

7. Make similar estimates to that of the previous question for a radio telescope with an aperture of 50 m operating at 30 cm wavelength.

8. State Rayleigh's criterion for the chromatic resolving power of a diffraction grating. The yellow sodium lines have wavelengths 589 nm and 589.6 nm. What is the minimum number of rulings N on a transmission grating for these lines to be resolved in first order ($m = 1$)? (UE)

9. The 643.8 nm cadmium line has a width of about 7×10^{-4} nm. How many rulings N must a diffraction grating have in order that the linewidth is resolvable in first order? In tenth order?

10. A plane wave travelling in the z direction falls on a diffracting screen in which there is a slit of varying transmissivity. At a distance x from the centre of the screen, a fraction $A(x)$ of the incident amplitude is transmitted. The screen is infinite in the y direction, and $A(x)$ is independent of y. Show that the amplitude of the diffracted wave in the direction θ is given by

$$u(\theta) = u_0 \exp(-i\omega t) \exp(ikL) \int_{-\infty}^{\infty} A(x) \exp(ikx \sin\theta)\, dx$$

where $k = 2\pi/\lambda$ is the wave number, u_0 is the incident amplitude, and $\exp(-i\omega t)$, $\exp(ikL)$ are appropriate phase factors. Find $u(\theta)$ if $A(x) = \exp(-u|x|)$. Sketch $u(\theta)$ and the intensity as functions of angle θ. (UE)

11. Find the blaze angle for scattering into 10th order at wavelength 10 μm for a grating with $D = 200$ μm.

12. If the blazed grating of the previous question is used at a wavelength of 5 μm, into which order is the light concentrated?

13. Find the interplane spacing d for the set marked c in figure 3.22.

14. Show that the three largest values of interplane spacing d for the simple cubic (rocksalt) structure in figure 3.25, are a, $a/\sqrt{2}$, $a/\sqrt{3}$.

15. Show that the general formula for interplane spacing in the square lattice of figure 3.22 is

$$d = a/(l^2 + m^2)^{1/2}$$

where l and m are integers.

16. Show that the general formula for interplane spacing in the simple cubic lattice is

$$d = a/(l^2 + m^2 + n^2)^{1/2}$$

where l, m and n are integers.

17. Use the result of the previous question to find (a) the three smallest, (b) the ten smallest Bragg scattering angles for X-rays of wavelength 0.028 nm scattering off rocksalt (a = 0.28 nm).

4

Waves and Particles

It has already been noted (section 2.8) that high frequency electromagnetic radiation can exhibit particle-like behaviour. For example, in the sensitive type of detector known as a photomultiplier tube (described in Wehr and Richards, *Physics of the Atom*, 2nd edn, Addison-Wesley, section 10.5), a beam of low-intensity X-rays is detected as a sequence of isolated sharp pulses, such as would be produced by a stream of particles arriving at the detector. As was pointed out, it is not surprising that it is usually high-frequency radiation which travels in particle form; as shown in figure 2.39, a wave group of high-frequency radiation can be compressed into a very short time span. However, when we speak of a particle, we mean something more than an entity which arrives in a short time span. A particle has a definite size, or more exactly it has a well-defined energy and momentum. It was discovered around about 1900 that the same is true of an X-ray wave packet: all packets of a given frequency radiation carry the same energy and momentum. This revolutionary discovery means that we must treat the wave packets as particles on exactly the same footing as more familiar particles such as electrons. The electromagnetic particle is called a *photon.* There is a wealth of experimental evidence which may be used to define the properties of a photon. We shall be highly selective, however, and simply present two particularly clear-cut experiments, the photoelectric effect and Compton scattering, which enable one to define the energy and momentum of the photon in terms of the frequency and wave number of the corresponding wave. These experiments will be described in detail in section 4.1.

The terminology used here is not exact, in principle. As was emphasised in chapter 1, because a wave packet occupies a finite time δt, it contains a natural spread of frequencies $\delta\omega_n \approx 2\pi/\delta t$. This natural spread was ignored when we spoke of relating the energy of a particle to the frequency of the corresponding wave packet. However, a wave packet of 10^{18} Hz X-rays occupying a time 10^{-9} s has a relative frequency spread $\delta\omega_n/\omega \approx 10^{-9}$. The X-rays used in the photoelectric effect and in Compton scattering are selected from a broad-band source by means of a Bragg crystal monochromator, as shown in figure 3.28. The X-rays diffracted into a cone of small angle $\delta\theta$ are selected, and this introduces an *instrumental bandwidth* $\delta\omega_i$ proportional to $\delta\theta$. It is easy to see that for any practical $\delta\theta$, $\delta\omega_i$ far exceeds $\delta\omega_n$. It is, therefore, reasonable to ignore the natural width $\delta\omega_n$ in discussing the experimental results, and to continue to use the simple terms frequency and wave number of a wave packet.

Once it had been established that electromagnetic radiation could manifest itself in particle form, it was natural to enquire whether particles might sometimes behave as waves. Once again we shall be selective, and in section 4.2 describe just one experiment, namely diffraction of electrons or neutrons by crystals, which shows that all particles do indeed have an additional wave-like character.

The systematic study of the relationship between wave and particle properties is known as *wave mechanics*, or *quantum mechanics.* This chapter goes only as far as setting out some of the basic notions.

Worked example 4.1

Calculate the relative instrumental bandwidth $\delta\omega_i/\omega$ for X-rays of frequency around 10^{18} Hz selected by taking the radiation scattered into an angle $\delta\theta = 10^{-2}$ radians around $\theta = \pi/12$ (15°) from a Bragg monochromator.

Answer

The Bragg scattering is governed by equation 3.38

$$2d \sin\theta = n\lambda$$

where d is the inter-plane spacing and n is the order of scattering. We can differentiate this relation to relate $\delta\theta$ to the bandwidth $\delta\lambda$

$$2d \cos\theta\,\delta\theta = n\,\delta\lambda$$

or

$$\delta\lambda = \frac{2d}{n}\cos\theta\,\delta\theta$$

Now, as in worked example 3.2, we have

$$\frac{|\delta f|}{f} = \frac{|\delta\lambda|}{\lambda}$$

This gives

$$\frac{|\delta f|}{f} = \frac{|\delta\lambda|}{\lambda} = \cot\theta\,\delta\theta$$

since

$$\lambda = \frac{2d}{n} \sin \theta$$

Since $\cot(\pi/12) = \sqrt{3} = 1.73$, we have

$$\frac{|\delta\omega_i|}{\omega} = \frac{|\delta f|}{f} = 1.73 \times 10^{-2}$$

4.1 Photons

The energy of a photon is found from the photoelectric effect. Monochromatic radiation, typically from the ultra-violet part of the spectrum, is allowed to fall on a clean metal surface. It is found that if the angular frequency ω of the radiation exceeds a certain threshold value ω_0, then electrons are ejected from the metal surface. Furthermore, for a given value of ω, the ejected electrons all have the same kinetic energy, and the energy increases linearly with ω. The results are summarised in figure 4.1.

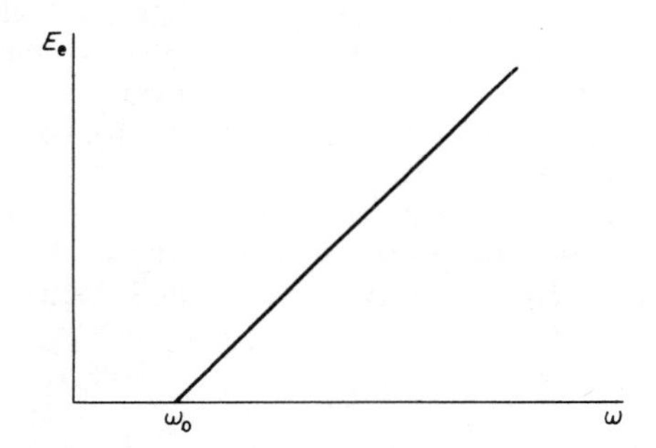

Figure 4.1 The photoelectric effect. The energy E_e of the electrons produced at a metal surface increases linearly with frequency ω of the incident radiation

The explanation of the photoelectric effect was given by Einstein. As shown in figure 4.2, an electron inside the metal has a lower potential energy than one outside the metal, since the electron is bound inside the metal. Now suppose we assume that the incident light arrives in the form of photons of energy E proportional to the frequency

$$E = \hbar\omega \qquad (4.1)$$

The constant of proportionality is of course unknown at this stage. If a photon is absorbed in the metal, its energy goes partly in raising an electron to the outside potential level, and partly in giving it kinetic energy E_e.

Conservation of energy requires

$$\hbar\omega = E_e + W \qquad (4.2)$$

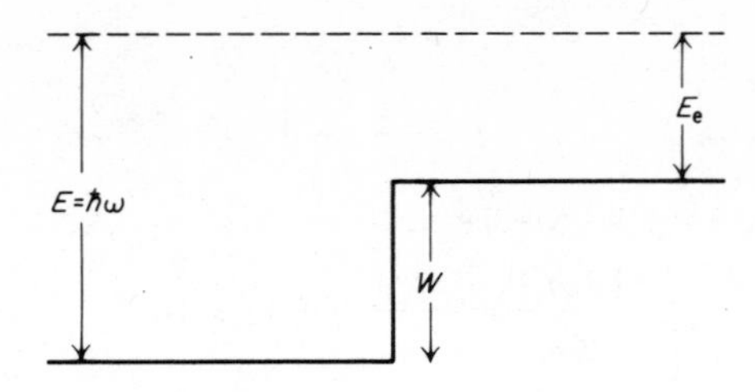

Figure 4.2 Illustration of the derivation of the equation for the photoelectric effect. Solid line: potential of electron inside and outside metal. Dotted line: energy of incident photon relative to inside potential

This is exactly the equation required to describe experimental results such as those of figure 4.1. If we write

$$W = \hbar\omega_0 \tag{4.3}$$

then equation 4.2 can be reorganised as

$$E_e = \hbar(\omega - \omega_0) \tag{4.4}$$

which is the equation of a straight line like that in figure 4.1. Equation 4.3 relates the threshold frequency ω_0 to W, which is characteristic of the particular metal being used, and is called the *work function* of the metal.

We have now seen that the assumption that the photons all have the same energy, given by equation 4.1, gives a simple explanation of the photoelectric effect, and indeed no other interpretation of the effect is possible. According to equation 4.4, the constant of proportionality, $\hbar$, is the slope of the line in figure 4.1, so the photoelectric effect is a way of measuring $\hbar$. Equation 4.1 is often written in terms of the ordinary frequency f as

$$E = hf \tag{4.5}$$

where h is called *Planck's constant.* Obviously

$$\hbar = h/2\pi \tag{4.6}$$

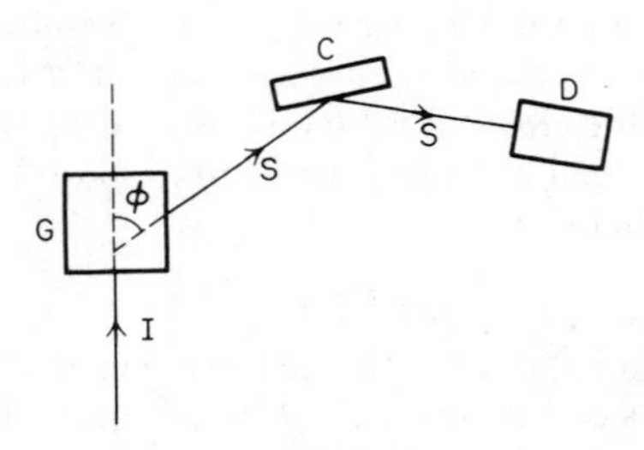

Figure 4.3 Compton scattering apparatus. The incident X-rays I, scatter off the graphite target G. The scattered X-rays, S, are analysed by the X-ray spectrometer, consisting of the crystal C and detector D

The value of h is 6.626×10^{-34} J s. In terms of the electron volt or eV (the energy acquired by an electron which falls through an electric potential difference of one volt) the value of h is 4.14×10^{-15} eV s. Equation 4.3 relates the work function W to the threshold frequency ω_0, and the numerical value we have quoted means that if ω_0 is in the ultra-violet, W is of the order of 10 eV. A few electron volts is the sort of value generally found for energy differences between

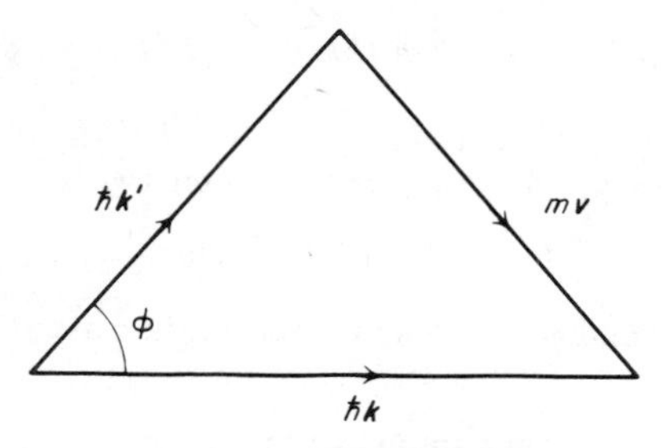

Figure 4.4 Momentum conservation triangle for Compton effect. The incident photon is assumed to have momentum $\hbar \boldsymbol{k}$, and the scattered photon, $\hbar \boldsymbol{k}'$. The difference of the two is the electron recoil momentum $m\boldsymbol{v}$

electrons inside and outside a material. This is not surprising, since the volt was defined in the first place in terms of the potential produced when electrons are moved between materials in a standard cell.

We now turn to the Compton effect, which gives the momentum of the photon. The experimental arrangement is shown in figure 4.3. Monochromatic X-rays of wavelength λ fall on a graphite target, and the scattered radiation is analysed with a Bragg X-ray spectrometer, as described in section 3.5. It is found that the scattered radiation has a longer wavelength λ', related to λ by

$$\lambda' - \lambda = (1 - \cos\phi)\frac{h}{m_0 c} \tag{4.7}$$

where ϕ is the angle between the incident and scattered X-rays, and m_0 is the rest mass of the electron. The multiplying factor h/m_0c has a value of 2.4 pm (2.4×10^{-12} m) so that for X-rays with a wavelength of around 1 nm, the shift is small compared with the incident wavelength. The explanation of the shift in wavelength was given by Compton and Debye. The incident X-rays, once again, must be regarded as a stream of photons. Some of the photons undergo collisions with electrons in the graphite as a result of which they lose energy and momentum, and it is these photons which emerge as the scattered radiation. We shall now see that we can derive equation 4.7 if we ascribe to the photon a momentum $\boldsymbol{p}$ proportional to the wave number $\boldsymbol{k}$

$$\boldsymbol{p} = \hbar \boldsymbol{k} \tag{4.8}$$

with the same constant of proportionality as in equation 4.1 for the energy.

We shall analyse the collision between a photon and an electron by ordinary mechanics, but we shall follow the usual custom of using relativistic expressions for the momentum and energy of the electron. In fact, if ordinary, newtonian expressions were used we should get a result differing from equation 4.7 only by terms of order v^2/c^2 where v is the recoil velocity of the electron. Since $v \ll c$, that result would be experimentally indistinguishable from equation 4.7. The derivation using the newtonian expressions is left for problem 4.4.

It can be seen from figure 4.4 that the triangle for conservation of momentum gives

$$m^2v^2 = (\hbar k)^2 + (\hbar k')^2 - 2\hbar^2 kk' \cos\phi \tag{4.9}$$

where we are assuming that the momentum vector of the photon is given by equation 4.8. The equation for conservation of energy is

$$\hbar\omega + m_0c^2 = \hbar\omega' + mc^2 \tag{4.10}$$

In both equations 4.9 and 4.10, m is the relativistic mass at speed v, related to the rest mass m_0 by

$$m = m_0(1 - v^2/c^2)^{-1/2} \tag{4.11}$$

Equation 4.7 is a relationship between the experimentally measured quantities λ, λ' and $\cos\phi$, and does not involve the electron recoil velocity v. In order to derive equation 4.7, then, we eliminate v between equations 4.9 and 4.10. First we replace the frequencies ω and ω' by the wave numbers k and k' in equation 4.10:

$$mc^2 = \hbar c(k - k') + m_0c^2 \tag{4.12}$$

We now square both sides of this equation in order to use equation 4.11 for the mass m

$$\frac{m_0^2c^2}{1 - v^2/c^2} = \hbar^2(k - k')^2 + 2\hbar c m_0(k - k') + m_0^2c^2 \tag{4.13}$$

where we have dropped a factor c^2 on either side. Subtracting $m_0^2c^2$ from either side of this equation gives

$$\frac{m_0^2v^2}{1 - v^2/c^2} = \hbar^2(k - k')^2 + 2\hbar c m_0(k - k') \tag{4.14}$$

The left-hand side of this equation is the same as the left-hand side of equation 4.9, since the latter involves the relativistic mass m of equation 4.11. Subtracting equation 4.14 from equation 4.9, thus eliminating v, we find

$$k - k' = \frac{\hbar}{m_0c} kk'(1 - \cos\phi) \tag{4.15}$$

It can be seen that this equation is identical to equation 4.7 if it is divided by kk' and the relations $\lambda = 2\pi/k$, $\lambda' = 2\pi/k'$, $h = 2\pi\hbar$ are used. Thus we have proved, following Compton and Debye, that the shift in wavelength can be explained on the assumption that the X-rays travel as a stream of photons, each having energy $\hbar\omega$ and momentum $\hbar k$.

For photons, the questions posed at the beginning of this chapter have now been answered. It was pointed out that for a wave packet to qualify as a particle, it must have a well-defined energy and momentum, and that we might expect these to be related to the frequency and wave vector of the light. The photoelectric effect and the Compton effect, taken together, show that the energy is proportional to the frequency and the momentum to the wave number, equations 4.1 and 4.8 respectively. The constant of proportionality, $\hbar$, measures the size of the photon, or individual *quantum* of energy. Equation 4.8, showing that the momentum is proportional to $\boldsymbol{k}$, deserves a little more discussion. We saw in section 3.2 that in two or three dimensions $\boldsymbol{k}$ is a vector of magnitude $2\pi/\lambda$, and the wave travels in the direction of the vector $\boldsymbol{k}$. Equation 4.8, therefore, has the satisfactory property that the momentum is in the direction of travel of the wave. We may rewrite it as an equation between the magnitude of $\boldsymbol{p}$, $p = |\boldsymbol{p}|$, and the wavelength

$$p = h/\lambda \tag{4.16}$$

Worked example 4.2

Estimate the electron recoil velocity v in 90° Compton scattering of 10^{18} Hz X-rays.

Answer

The recoil velocity is given by equation 4.14, with the wave number shift of equation 4.15. Since $k = 2\pi/\lambda$, we have for a small shift of wave number,

$$\frac{|\delta k|}{k} = \frac{|\delta\lambda|}{\lambda}$$

For 90° scattering, equation 4.7 gives $\delta\lambda = h/m_0c = 2.4$ pm which is small compared with the wavelength $\lambda = 3 \times 10^8/10^{18} = 300$ pm. This means that $\delta k = k - k'$ is small compared with k, so we may neglect the difference between k and k' on the right in equation 4.15. Equation 4.15 then reads, for 90° scattering,

$$k - k' = \frac{\hbar}{m_0 c} k^2$$

The numerical value of k is $2\pi/\lambda = 2\pi/(3 \times 10^{-10}) = 2.1 \times 10^{10}\ \text{m}^{-1}$. The shift is therefore

$$k - k' = \frac{2.4 \times 10^{-12}}{2\pi} \times (2.1 \times 10^{10})^2\ \text{m}^{-1}$$

We assume, tentatively, that the electron recoil velocity v is non-relativistic, $v \ll c$. Equation 4.14 then simplifies to

$$m_0{}^2 v^2 = \hbar^2(k - k')^2 + 2\hbar m_0 c(k - k')$$

Putting in the numerical values, we find that the second term is dominant, and gives

$$v = 3.4 \times 10^6 \text{ m s}^{-1}$$

Thus $v \approx 10^{-2}\ c$, and the assumption that $v \ll c$ is justified.

4.2 Wave Character of Particles

After the nature of photons had been clarified, it was suggested by de Broglie that more familiar particles, such as electrons, might be associated with a wave motion. He proposed further that the wave number and wavelength might be related to

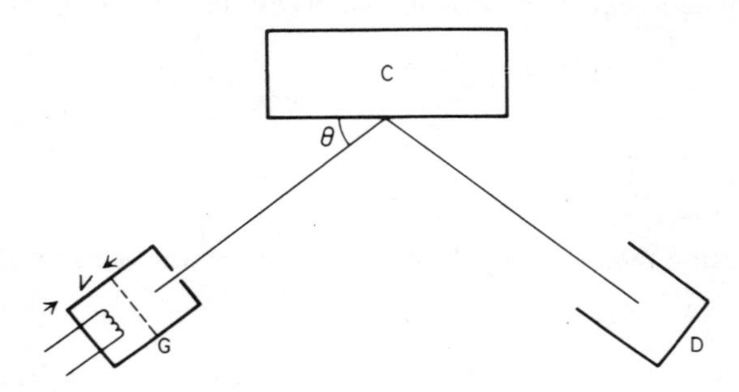

Figure 4.5 The Davisson and Germer experiment. Electrons accelerated through a voltage V in the gun G are diffracted off the nickel crystal C and detected at D. The diffraction is measured as a function of V and θ

the momentum of the particle by the same equations as hold for photons, equations 4.8 and 4.16. This proposal was confirmed by Davisson and Germer, who showed that an electron beam could undergo Bragg diffraction by a single crystal. As shown in figure 4.5, they scattered electrons of known energy, and therefore, of known momentum, off a single crystal of nickel. It was found that strong reflections occurred at particular angles θ, just as in Bragg diffraction of X-rays. The lattice structure of nickel had already been determined by X-ray diffraction, so the plane spacings d were known. The Bragg scattering condition, equation 3.38, could then be used to find the wavelength of the electrons from the scattering angle. It was found, indeed, that equation 4.16 holds, as proposed by de Broglie.

We have already seen (section 2.6) that tunnelling of electrons and α-particles implies that they have a certain wave character. The electron diffraction experiment is more precise, however, in that it determines the wavelength. It has now been established by diffraction experiments that all particles have an associated wave, and that equations 4.8 and 4.16 are universal relations between the momentum of the particle and the properties of the associated wave. The 'displacement' associated with the wave is called the *wave function* and the interpretation of the wave function will be dealt with in the next section.

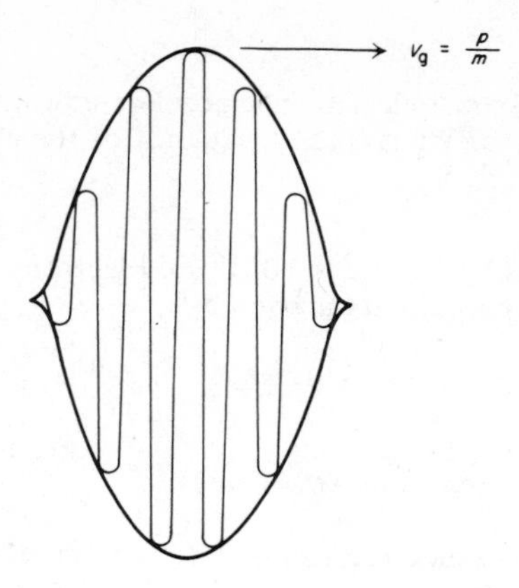

Figure 4.6 Group velocity must equal particle velocity

We still lack, for electrons, a connection between energy and frequency, but it can easily be seen that equation 4.1 holds for electrons as well as for photons. It is necessary to regard the electrons as a wave packet of the associated wave (figure 4.6), and in particular this means that the group velocity at which the wave packet travels must be equal to the ordinary particle velocity p/m. Thus we have

$$\frac{d\omega}{dk} = \frac{p}{m} \tag{4.17}$$

or using equation 4.8

$$\frac{d\omega}{dp} = \frac{p}{\hbar m} \tag{4.18}$$

Integrating this equation, we get

$$\omega = \frac{1}{\hbar} \frac{p^2}{2m} \tag{4.19}$$

Comparison with the value of the energy

$$E = \tfrac{1}{2} mv^2 = \frac{p^2}{2m} \tag{4.20}$$

gives indeed

$$E = \hbar\omega \tag{4.21}$$

which is the same as equation 4.1. This equation is therefore universal for photons and particles.

Worked example 4.3

Through what potential must electrons be accelerated to have a wavelength equal to that of 10^{18} Hz X-rays? What is the frequency of the electrons?

Answer

The X-ray wavelength is $\lambda = c/f = 3 \times 10^{-10}$ m. Electrons accelerated through a potential of V volts have a velocity v given by

$$\tfrac{1}{2} mv^2 = eV$$

or

$$v = (2eV/m)^{1/2}$$

assuming $v \ll c$ so that the electron can be treated non-relativistically. The corresponding momentum is

$$p = mv = (2emV)^{1/2}$$

and the wavelength λ_e of the electron is

$$\lambda_e = h/p$$

We therefore have

$$h/(2emV)^{1/2} = \lambda$$

where λ is the X-ray wavelength. This gives

$$V = \frac{h^2}{2em\lambda^2}$$

or putting in the numbers, $V = 17.2$ volts. The corresponding velocity, $v = (2eV/m)^{1/2}$ is 2.5×10^6 m s^{-1} which confirms the assumption $v \ll c$. The frequency, from equation 4.5, is $f = eV/h = 4.2 \times 10^{15}$ Hz.

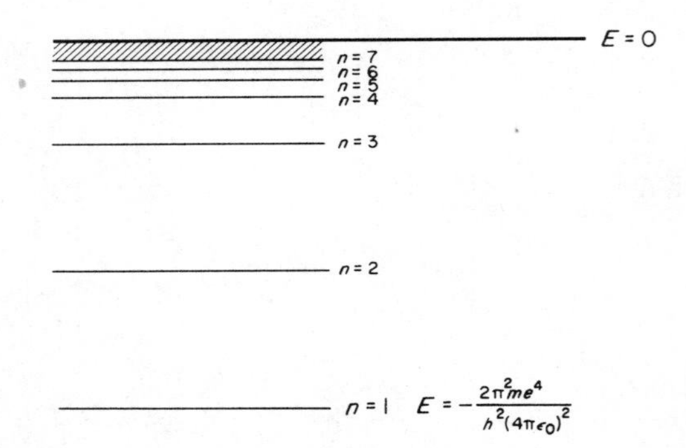

Figure 4.7 Energy-level scheme for hydrogen. $E = 0$ is taken as the energy of an electron at infinity relative to the proton

4.3 Particle Waves and Energy Levels

It is of interest to see how the wave character of electrons gives a natural explanation of the existence of quantised energy levels in atoms. It will be recalled that atoms emit characteristic line spectra; that is, the radiation emitted by an element such as neon in a discharge tube consists of a set of separated lines each of essentially one frequency. Each line is associated with the transition of an electron from one sharply defined energy level within the atom to another. The energy levels are described by quantum numbers; for example, the energy level scheme of hydrogen is

$$E_n = -\frac{2\pi^2 me^4}{h^2(4\pi\epsilon_0)^2}\frac{1}{n^2} \tag{4.22}$$

where n is an integer, called the *principal quantum number.* This energy level scheme is shown in figure 4.7. The various frequencies ω of the spectral lines are given by the general relation

$$\hbar\omega = E_n - E_m \tag{4.23}$$

where E_n and E_m are two different levels with $E_n > E_m$. Equation 4.23 shows simply that when an electron drops from level n to level m the energy released appears as a photon. It is found that in all atoms the energy levels are related to integers by formulae which are similar to, although generally more complicated than, equation 4.22.

At first, formulae like equation 4.22 were derived simply as empirical formulae which fitted the line spectra. We can see, however, that equations relating energy levels to integers arise naturally from a standing-wave condition on the wave associated with the electron. The potential energy of an electron in hydrogen is sketched in figure 4.8 as a function of the radial distance r of the electron from the proton which is the hydrogen nucleus. If the electron were simply a particle,

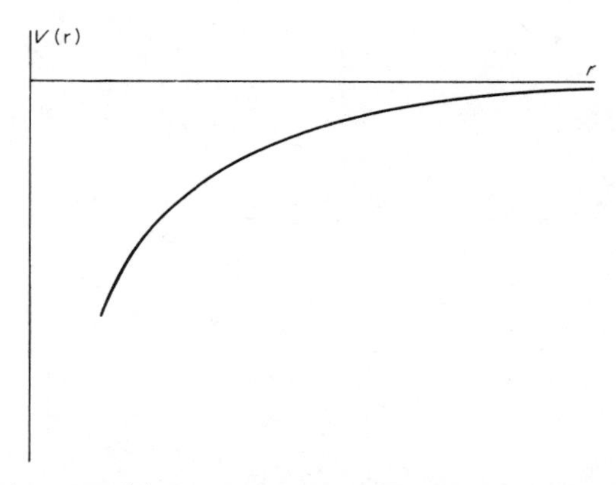

Figure 4.8 Potential energy $V(r)$ of an electron at a distance r from a proton. The potential energy is simply the electrostatic energy, $V(r) = -e^2/(4\pi\epsilon_0\gamma)$

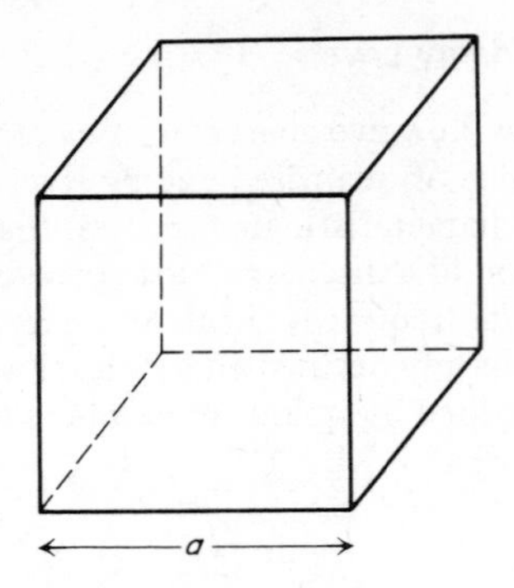

Figure 4.9 A particle box

a bound state would correspond to motion of the electron in an orbit round the proton, like the orbit of the earth round the sun. The corresponding wave is clearly localised round the proton. However, to localise a wave we must impose some boundary condition on the wave, in the present case the condition that the wave becomes small as $r \to \infty$. Because of the boundary condition, it turns out that only certain orbits are possible, and equation 4.22 gives the energies corresponding to the allowed orbits.

In order to describe a particle wave in the presence of a varying potential, like that of figure 4.8, we need to use what is called the *Schrödinger equation*. The Schrödinger equation will be dealt with in chapter 6, and it will be used to derive the hydrogen ground-state energy E_1. For the present, however, we can use a simple system to see how localisation of a particle requires a standing-wave condition which in turn leads to separated, or discrete, energy levels. Consider a

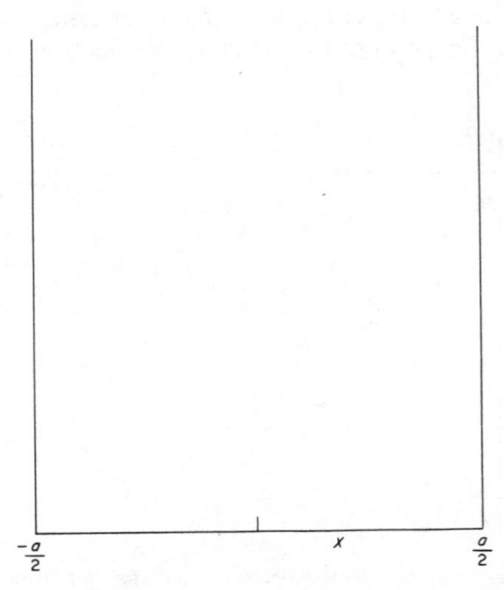

Figure 4.10 The potential energy is taken as 0 for $-a/2 < x < a/2$, and ∞ outside that range

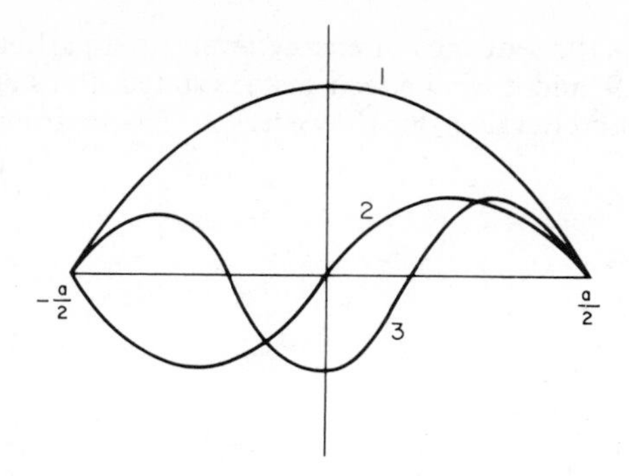

Figure 4.11 First three standing waves in the potential well of figure 4.10

particle moving in a cubic box of side a, as shown in figure 4.9. It is assumed that the potential energy is zero inside the box, and infinite outside it, so that the particle and the associated wave function cannot penetrate outside the box. The cross section of the potential energy in the x direction is shown in figure 4.10. The wave function must vanish outside the box, which means that the wave function inside the box must be a standing wave that vanishes at the boundaries. The x dependence of the wave function for the first three standing waves is shown in figure 4.11. The explicit forms are

$$u_1(x) = \cos k_1 x \qquad \text{with } k_1 a/2 = \pi/2 \tag{4.24}$$

$$u_2(x) = \sin k_2 x \qquad \text{with } k_2 a/2 = \pi \tag{4.25}$$

$$u_3(x) = \cos k_3 x \qquad \text{with } k_3 a/2 = 3\pi/2 \tag{4.26}$$

We see, therefore, that because of the boundary conditions there is a sequence of *allowed values* of the wave number k_x; that is k_x must be an integral multiple of π/a

$$k_x = l\pi/a \tag{4.27}$$

where l is an integer. Similarly, k_y and k_z must be integral multiples of π/a

$$k_y = m\pi/a \tag{4.28}$$

$$k_z = n\pi/a \tag{4.29}$$

We can easily derive an expression for the energy corresponding to a given standing wave from equations 4.27 to 4.29. The energy is

$$E = (p_x^2 + p_y^2 + p_z^2)/2m_e \tag{4.30}$$

and since each component of momentum is related to the corresponding wave number by equation 4.8

$$E = \frac{\hbar^2\pi^2}{2m_e a^2}(l^2 + m^2 + n^2) \tag{4.31}$$

Equation 4.31 gives the sequence of energy levels for a particle localised in the cubic box of figure 4.9, and it can be seen that, as stated, the standing-wave conditions result in discrete values for the energy E. The first few values of the

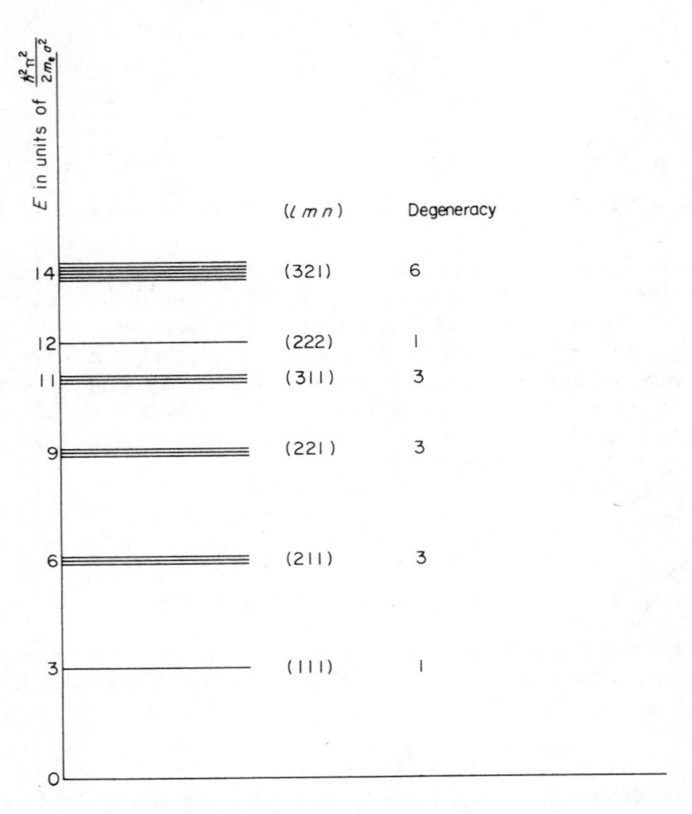

Figure 4.12 Energy-level scheme for the cubic box

energy are sketched in figure 4.12, which is the equivalent for the present problem of the hydrogen level scheme shown in figure 4.7. It will be seen first of all that the ground-state energy, corresponding to $l = m = n = 1$, is not zero. A classical particle would come to rest at any point in the box with energy zero; but because of the wave nature of the electron the actual lowest energy is greater than zero. This *zero-point energy* will be looked at from a different point of view in section 4.5. A second feature of the energy level scheme is that many of the energy levels are *degenerate.* For example, the combination $l^2 + m^2 + n^2 = 6$, contains the three different states $l = 2, m = 1, n = 1; l = 1, m = 2, n = 1; l = 1, m = 1, n = 2$. The degeneracy of each level is shown in figure 4.12.

Of the two features we have emphasised, zero-point energy occurs in all energy level schemes, and degeneracy in most. In the hydrogen level scheme, for example, the ground state, $n = 1$, includes some zero-point energy, since a classical particle

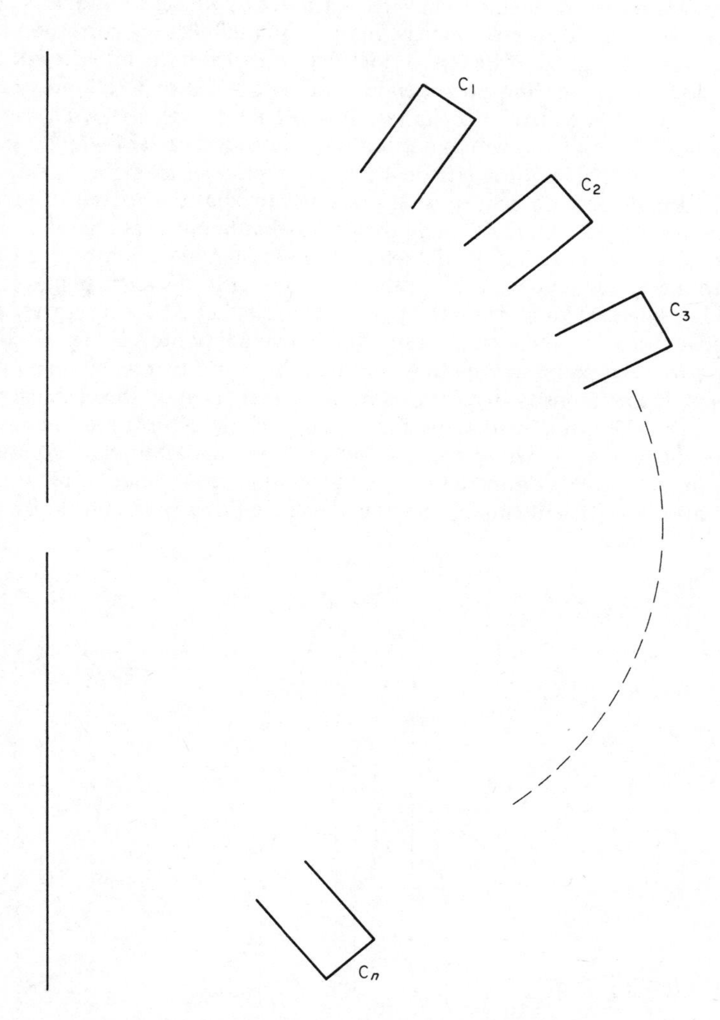

Figure 4.13 Single-slit diffraction with a battery of counters

in the potential of figure 4.8 would collapse to the origin with energy $-\infty$. The levels above the ground state in hydrogen are in fact degenerate; the nth level having degeneracy n^2.

4.4 Interpretation of the Wave Function

So far, we have spoken of the wave 'corresponding to' an electron, or of the photons 'described by' an electromagnetic wave, but we have not discussed precisely what the correspondence is. In order to establish the correspondence, consider the diffraction of light at a single slit; classically the intensity of the diffracted light has the angular dependence shown in figure 3.7. Suppose the usual screen, on which the diffraction pattern is observed, is replaced by an angular distribution of counters, such as photomultiplier tubes, each of which can detect the arrival of a single photon (figure 4.13). Furthermore, suppose that the experiment is carried out with light of such low intensity that the arrivals of individual photons are well separated. We may then ask, what happens as one photon passes the slit? The photon cannot be detected by more than one counter, for example it cannot be spread across all the counters according to the angular pattern of figure 3.7, since we know from the photoelectric effect that what arrives at a counter is either one photon, or none. The individual photon is therefore detected by one particular counter. We know on the other hand that when many photons have arrived the intensity distribution must be that given by the classical pattern of figure 3.7. Thus if a histogram of counts in each detector is plotted against angular position, as shown in figure 4.14, we know that after many counts the histogram will agree with the classical pattern. Since each detector records some counts after a sufficient time, it must be concluded that we do not know at which

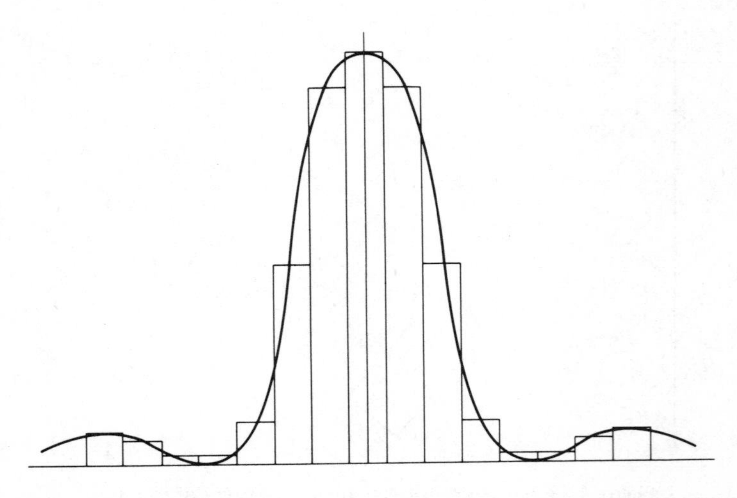

Figure 4.14 Histogram for counting experiment of figure 4.13. Vertical axis: number of counts. Horizontal axis: angular position

detector an individual photon will arrive. All that can be said is that the diffraction curve in figure 4.14 gives the *probability* that the photon will arrive at a given counter, since the counts eventually build up to the curve. The intensity shown in figure 3.7 is proportional to the square of the wave amplitude. Therefore for photons it must be concluded that the square of the wave amplitude at a point of the screen gives the probability that a photon will arrive at that point of the screen.

The diffraction of electrons or other 'particles' at a single slit gives an intensity distribution which is identical to that found for photons. The interpretation arrived at for photons is therefore quite general: the square of the wave function at a point gives the probability that a particle will be detected at that point.

4.5 The Uncertainty Principle

It was stressed in chapter 1 and subsequently that the range $\delta\omega$ of frequencies within a wave packet is related to the duration δt by

$$\delta\omega\delta t \approx 2\pi \tag{4.32}$$

Similarly the spread of wave numbers δk is given by

$$\delta k\delta x \approx 2\pi \tag{4.33}$$

where δx is the length of the wave packet in space. These relations apply equally to photons and other particles, which travel as wave packets of the corresponding

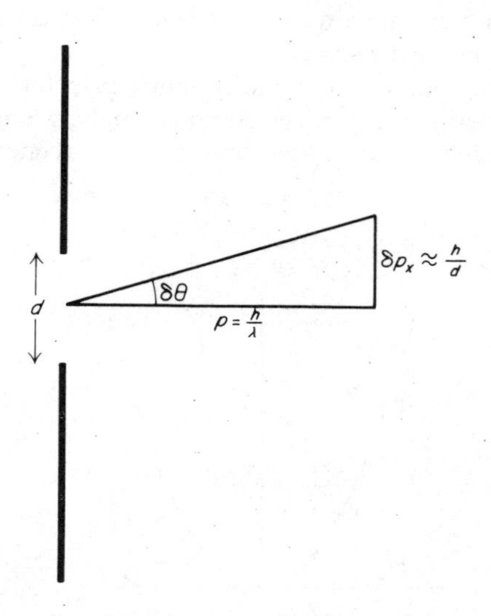

Figure 4.15 Diffraction from the point of view of the uncertainty principle

wave. With the aid of equation 4.1 for energy E and equation 4.8 for momentum p we may rewrite them as

$$\delta E \delta t \approx h \tag{4.34}$$

$$\delta p \delta x \approx h \tag{4.35}$$

where $h = 2\pi\hbar$ as usual. Equations 4.34 and 4.35 are known as the *quantum mechanical uncertainty relations*, or simply the *uncertainty principle.* They are derived here as a consequence of the 'classical' equations 4.32 and 4.33. Historically in fact, uncertainty relations were derived for quantum mechanics, and then taken over for classical waves.

One form of application of the uncertainty principle is similar to the kind with which we have become familiar. For diffraction of electrons at a single slit, for example, the restriction $\delta x \approx d$ at the slit implies a momentum spread $\delta p_x \approx h/d$. In a small-angle approximation, the angular spread of the diffracted beam, as shown in figure 4.15, is $\delta\theta \approx \delta p_x/p$ where p is the total momentum. Since p is related to the wavelength λ by equation 4.16, we find finally

$$\delta\theta \approx \delta p_x/p \approx \lambda/d \tag{4.36}$$

The argument used here, and the conclusion, are identical to those in section 3.2.

A second use of the uncertainty principle is to emphasise that quantum mechanics imposes a fundamental restriction on measurement. Equation 4.35, in particular, shows that if in some ideal experiment the position of a particle is measured exactly with no uncertainty δx, then the momentum is completely unknown, since we then have $\delta p = \infty$. Conversely, if the momentum is measured exactly, then the position is unknown. Any determination of momentum and position is limited to uncertainties δp and δx related by equation 4.35. More detailed discussions of the problem of measurement in quantum mechanics are given in most textbooks on the subject.

Finally, one may sometimes use the uncertainty principle to find an estimate of the ground-state energy of a particle. As an example, we may consider a particle moving in one dimension in such a way that its potential energy at the point x is

$$V(x) = \tfrac{1}{2}Kx^2 \tag{4.37}$$

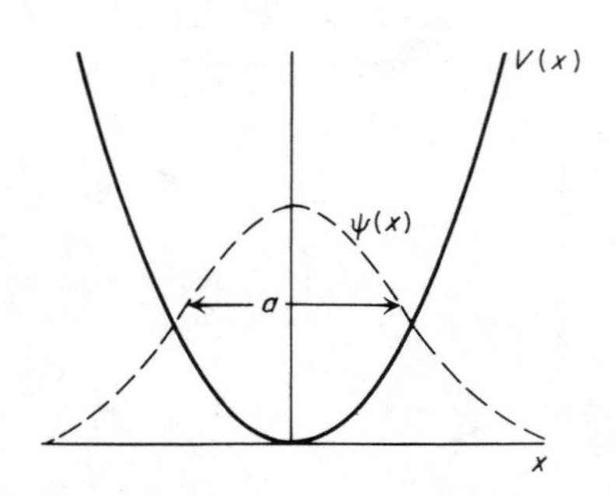

Figure 4.16 Harmonic oscillator potential and ground-state wave function

as sketched in figure 4.16. We shall see in chapter 5 that this is the potential energy of a particle undergoing simple harmonic motion. Suppose the particle in its ground state is described by a wave function $\psi(x)$ having a range a, as also sketched in figure 4.16. This spread implies a potential energy for the particle of order

$$V \approx \tfrac{1}{2}Ka^2 \tag{4.38}$$

The uncertainty principle implies a spread in momentum of order h/a, and since the kinetic energy T is $p^2/2m$, this gives

$$T \approx \frac{h^2}{2ma^2} \tag{4.39}$$

The total energy is therefore

$$E = T + V \approx \frac{h^2}{2ma^2} + \tfrac{1}{2}Ka^2 \tag{4.40}$$

T, V and E are sketched in figure 4.17 as functions of the degree of localisation a. If a is small, the momentum spread is small and the kinetic energy T dominates. Conversely, if a is large, the potential energy V is large. The minimum value of E

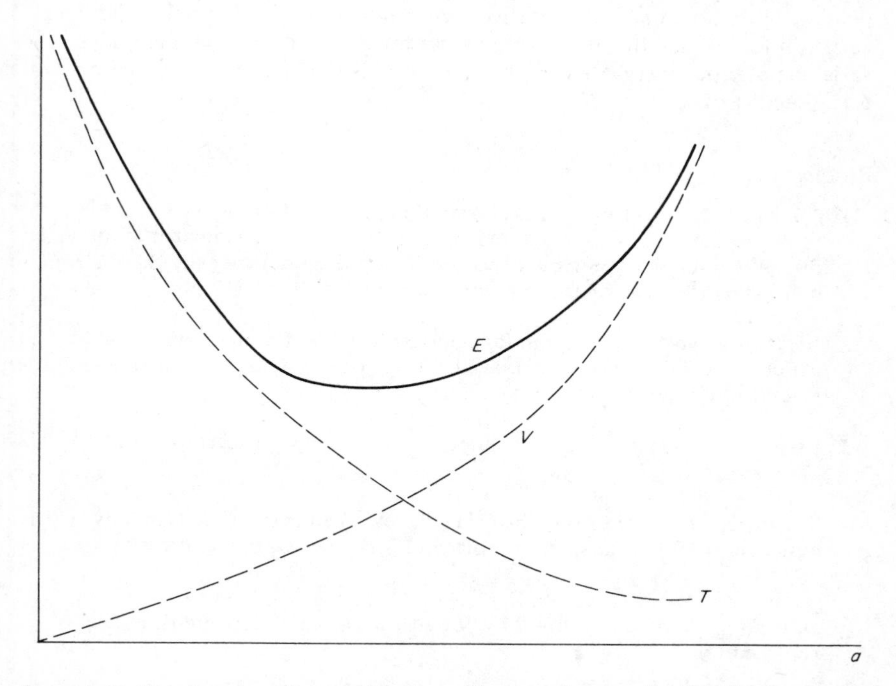

Figure 4.17 Kinetic energy T, potential energy V and total energy $E = T + V$ as functions of a for the wave function of figure 4.16.

as a function of a gives an estimate of the ground-state energy. The equation for a minimum is

$$\frac{\mathrm{d}E}{\mathrm{d}a} = -\frac{h^2}{ma^3} + Ka = 0 \tag{4.41}$$

which gives for the value of a at the minimum

$$a^4 = \frac{h^2}{mK} \tag{4.42}$$

and for the energy itself

$$E \approx h\left(\frac{K}{m}\right)^{1/2} \tag{4.43}$$

This is an estimate of the zero-point energy of the particle, which may be compared with the exact value of $h/4\pi(K/m)^{1/2}$ that will be derived in chapter 6. We can see now that the zero-point energy arises from the spreading of the wave function which is required by the uncertainty principle. The wave function cannot be concentrated at the point of zero potential, like a classical particle, since that would imply a large momentum spread and a large kinetic energy.

We have now pursued the study of quantum mechanics about as far as is possible with more or less qualitative arguments. To go further will require an equation to describe the exact form of the wave function in the presence of a variable potential energy. The appropriate tool is the Schrödinger equation, which is the subject of section 6.7.

Problems

1. Consider photons (wave packets) of X-rays each of duration $\tau = 10^{-12}$ s. Suppose the X-rays are scattered from a Bragg monochromator at an angle of 30°. What angular aperture $\delta\theta$ would be required in order for the instrumental bandwidth $\delta\omega_i$ to equal the natural bandwidth $2\pi/\tau$?

2. Light of wavelength 420 nm falls on the surface of a metal with a work function of 2.1 V. Find the kinetic energy of the emitted electrons, and their velocity, momentum and wavelength.

3. Estimate the electron recoil velocity in 60° Compton scattering of 10^{17} Hz X-rays.

4. Show that if in Compton scattering the electron is treated classically, with equation 4.10 for the energy replaced by the newtonian expression

$$\hbar\omega = \hbar\omega' + \tfrac{1}{2}m_0v^2$$

and the mass in equation 4.9 by the rest mass m_0, then equation 4.15 is replaced by

$$k - k' = \frac{\hbar}{2m_0c}(k^2 + k'^2 - 2kk'\cos\phi)$$

5. Show that the difference between the newtonian result for $k - k'$ derived in the previous example, and the relativistic result of equation 4.15 is

$$\Delta(k - k') = \frac{\hbar}{2m_0c}(k - k')^2$$

Using the numerical values of worked example 4.2, show that this difference is negligible compared with the value of $k - k'$ derived from either equation.

6. Find the accelerating potential to give a wavelength equal to that of 10^{18} Hz X-rays for (a) protons (b) α-particles (charge $2e$ mass $4\,m_p$).

7. What are the particle momenta in the previous question?

8. A particle moves in two dimensions in a square box of side a. Derive the formula, corresponding to equation 4.31, for the energy levels. Make a sketch, corresponding to figure 4.12, to show the first few levels with their degeneracies.

9. Use the uncertainty principle to estimate the zero-point energy of (a) a proton (b) an α-particle inside a nucleus. (Nuclear radius = 1 fm.)

10. Show, using the uncertainty principle, that an electron can be a constituent part of an atom, but is unlikely to be a constituent part of a nucleus. (Atomic radius = 1 nm; atomic binding energy = 10 eV per particle; nuclear radius = 1 fm = 10^{-15} m; nuclear binding energy = 8 MeV per particle; 1 eV = 1.6×10^{-19} J.) Compare the zero-point energy of the electron, derived from the uncertainty principle, with the binding energy per particle. (UE)

11. A particle has potential energy $\frac{1}{2}Kx^4$. Use the uncertainty principle, as in section 4.5, to estimate the ground-state energy.

12. A particle of mass m is confined to a rectangular box of sides a_1, a_2, a_3. Use the method of section 4.3 to find the allowed energy levels.

5

Resonance

Up to now, we have discussed the general properties of waves, mainly using the principle of superposition. In this chapter, we turn to the interaction of a wave with the medium in which it is travelling, and we shall concentrate on the medium itself. We shall find that the driven damped harmonic oscillator serves as a model for the response of any medium to an incident wave. The line of argument is as follows. First we show that in general the small vibrations about equilibrium of any system are simple harmonic oscillations. This gives the simple harmonic oscillator great general significance. A wave of frequency ω incident upon a system may then be represented by a force of frequency ω driving the harmonic oscillators corresponding to the small vibrations. In order to make our model physically realistic, we must allow for some damping of the vibrations by dissipative forces analogous to resistance in electrical circuits. The most important quantity to calculate is the power absorbed from the driving force by the oscillator. As a function of frequency, the power absorbed is a very characteristic *resonance curve*, or *lorentzian*, with a maximum at the frequency of the undamped oscillator. We shall finally look at a few of the many examples of resonance curves in natural systems.

5.1 Potential Functions and Small Vibrations

In this section, our first aim is to give a precise meaning to the idea of a potential energy curve for particle motion, which has already been used in section 2.7

(figure 2.37), and in section 4.5. We start with the equation of motion for the simple harmonic oscillator, equation 1.1

$$m\frac{\mathrm{d}^2u}{\mathrm{d}t^2} = -Ku \tag{5.1}$$

(Recall that u stands for a general displacement, such as the distance from the origin of a mass on a spring.) We can derive from equation 5.1 the important result that energy is conserved during the motion. We multiply equation 5.1 by the velocity $v = \mathrm{d}u/\mathrm{d}t$

$$m\frac{\mathrm{d}^2u}{\mathrm{d}t^2}\frac{\mathrm{d}u}{\mathrm{d}t} + Ku\frac{\mathrm{d}u}{\mathrm{d}t} = 0 \tag{5.2}$$

The second term simplifies, since

$$u\frac{\mathrm{d}u}{\mathrm{d}t} = \frac{\mathrm{d}}{\mathrm{d}t}(\tfrac{1}{2}u^2) \tag{5.3}$$

by the usual rule for differentiating a function of a function. Similarly, by writing

$$y = \frac{\mathrm{d}u}{\mathrm{d}t} \tag{5.4}$$

we see that the first term contains

$$y\frac{\mathrm{d}y}{\mathrm{d}t} = \frac{\mathrm{d}}{\mathrm{d}t}(\tfrac{1}{2}y^2) = \frac{\mathrm{d}}{\mathrm{d}t}\left[\frac{1}{2}\left(\frac{\mathrm{d}u}{\mathrm{d}t}\right)^2\right] \tag{5.5}$$

Thus equation 5.2 may be rewritten as

$$\frac{\mathrm{d}}{\mathrm{d}t}\left[\tfrac{1}{2}m\left(\frac{\mathrm{d}u}{\mathrm{d}t}\right)^2 + \tfrac{1}{2}Ku^2\right] = 0 \tag{5.6}$$

This means, finally, that the term in square brackets must be constant during the motion

$$\tfrac{1}{2}m\left(\frac{\mathrm{d}u}{\mathrm{d}t}\right)^2 + \tfrac{1}{2}Ku^2 = E \tag{5.7}$$

Equation 5.7 expresses the conservation of energy. The first term is simply the *kinetic energy* $\frac{1}{2}mv^2$. The second term is the *potential energy* V; for a mass on a spring, for example, it represents the energy stored in the spring when the displacement is u. It is convenient to look at the motion in terms of the potential energy function $V(u)$

$$V(u) = \tfrac{1}{2}Ku^2 \tag{5.8}$$

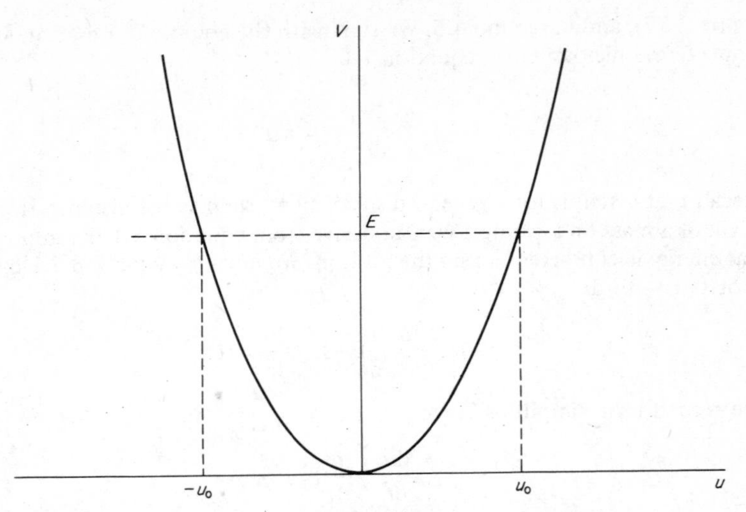

Figure 5.1 Potential energy for simple harmonic oscillator

As shown in figure 5.1, $V(u)$ is a parabola. Because of the shape of the curve, $V(u)$ is sometimes called the *potential well* for the motion. At the extreme points of the motion, the velocity $v = \mathrm{d}u/\mathrm{d}t$ is zero, and consequently the kinetic energy is zero. It can be seen from equation 5.7, therefore, that the extreme points are given by $u = \pm u_0$, with

$$\tfrac{1}{2}Ku_0^2 = E \tag{5.9}$$

as shown in figure 5.1. Conversely, at the origin, $u = 0$, the potential energy is zero, and equation 5.7 gives for the maximum velocity $v = \pm v_0$, with

$$\tfrac{1}{2}mv_0^2 = E \tag{5.10}$$

During the motion, then, the energy alternates between all potential at the extreme points and all kinetic at the origin, with the sum of kinetic and potential energies always constant.

It is now easy to generalise the definition of a potential function $V(u)$. If the motion is such that a potential function can be defined, then we must have conservation of energy

$$\tfrac{1}{2}m\left(\frac{\mathrm{d}u}{\mathrm{d}t}\right)^2 + V(u) = E \tag{5.11}$$

We cannot expect to find an equation of this kind for all motions; for example, any frictional force removes energy from the system, so that energy is not conserved. We must, therefore, ask what form must the equation of motion take in order that the *energy integral* of equation 5.11 can be derived? This question can be answered by reversing the steps that led from equation 5.1 to equation 5.7. In fact, by differentiating equation 5.11 with respect to time, we find

$$\frac{\mathrm{d}}{\mathrm{d}t}\left[\tfrac{1}{2}m\left(\frac{\mathrm{d}u}{\mathrm{d}t}\right)^2 + V(u)\right] = 0 \qquad (5.12)$$

which leads to

$$m\frac{\mathrm{d}^2u}{\mathrm{d}t^2}\frac{\mathrm{d}u}{\mathrm{d}t} + \frac{\mathrm{d}V}{\mathrm{d}u}\frac{\mathrm{d}u}{\mathrm{d}t} = 0 \qquad (5.13)$$

Cancelling $\mathrm{d}u/\mathrm{d}t$, the equation of motion becomes

$$m\frac{\mathrm{d}^2u}{\mathrm{d}t^2} = -\frac{\mathrm{d}V}{\mathrm{d}u} \qquad (5.14)$$

Since the left-hand side of this equation is mass times acceleration, the right-hand side must be the force. We conclude, therefore, that if the force F is given by

$$F = -\frac{\mathrm{d}V}{\mathrm{d}u} \qquad (5.15)$$

then energy is conserved as in equation 5.11.

Equation 5.15 gives the condition for which a potential function $V(u)$ can be defined. As was pointed out, not all forces can be written in this form. The simplest

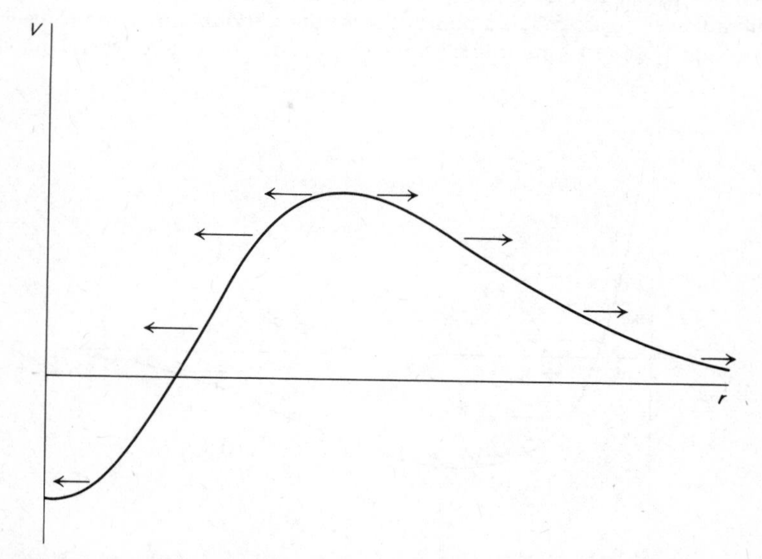

Figure 5.2 Potential energy of an α-particle as a function of distance from a nucleus. The arrows show the direction of the force, with the length of the arrow indicating the magnitude of the force at that point

form of frictional force, for example, is simply proportional to the velocity

$$F_{\mathrm{f}} = -b \frac{\mathrm{d}u}{\mathrm{d}t} \tag{5.16}$$

A force of this kind, which depends on velocity rather than displacement, cannot be written in terms of a potential.

Equation 5.15 gives a clearer understanding of potential energy diagrams like figure 5.1 for the harmonic oscillator, or figure 2.37 for the α-particle potential, since it shows that the force is proportional to the slope of the potential function, and is in such a direction as to decrease the potential. The direction of the force at various points of the α-particle potential is shown in figure 5.2. As another example figure 5.3 shows the form of potential energy curve which is typical for the atoms of a diatomic molecule, such as H_2. It can be seen from these two examples that a maximum or a minimum in the curve is a point of equilibrium of the particle, since the force is zero there. This follows directly from equation 5.15, since a zero force implies

$$\frac{\mathrm{d}V}{\mathrm{d}u} = 0 \tag{5.17}$$

which is the usual condition for a maximum or minimum in V. It can be seen, too, that a maximum, like that in figure 5.2, is a position of *unstable equilibrium.* If the particle moves to the left of the maximum the force moves it further to the left, and if it moves to the right, the force moves it further to the right. Conversely, a minimum, as in figure 5.3, is a point of *stable equilibrium*, since the force which appears is always a restoring force. Expressing the distinction between a maximum

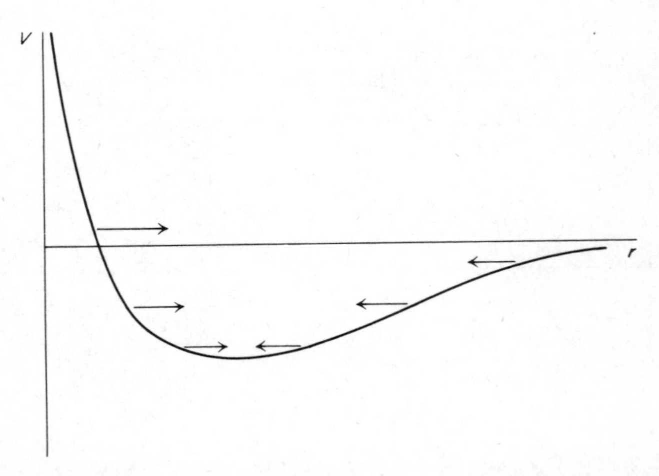

Figure 5.3 Potential energy as a function of separation of the atoms in a diatomic molecule. The force at various points is shown

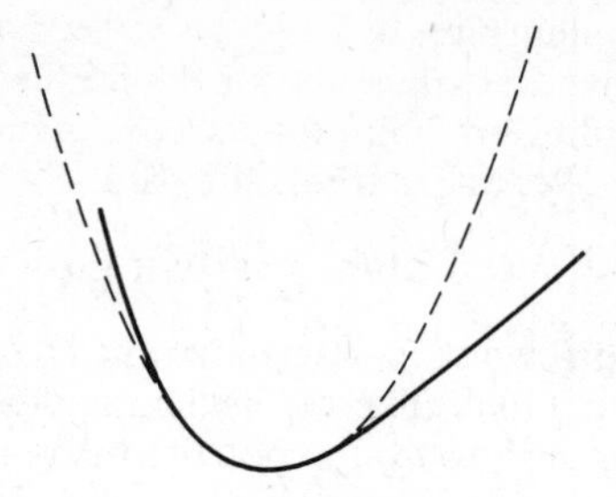

Figure 5.4 Near the minimum, the potential function of figure 5.3 coincides with a parabola (dashed curve)

and a minimum in the usual way in terms of the second derivative of V, gives the following rule for equilibrium points

$$\frac{dV}{du} = 0 \qquad \frac{d^2V}{du^2} < 0 \qquad \text{unstable equilibrium} \tag{5.18}$$

$$\frac{dV}{du} = 0 \qquad \frac{d^2V}{du^2} > 0 \qquad \text{stable equilibrium} \tag{5.19}$$

We may now investigate the small vibrations about a stable equilibrium point, like the minimum in figure 5.3. As sketched in figure 5.4, for small displacements from the minimum, the potential function does not depart much from a suitably chosen parabola. Equivalently, as shown in figure 5.5, the force on the particle is close to the linear restoring force of the harmonic oscillator. It may be concluded that for small enough displacements, the vibrations about equilibrium are the same as simple harmonic oscillations in the parabolic well shown dashed in figure 5.4.

The geometric argument which we have just given is sufficient to prove the main point of this section, that the simple harmonic oscillator is a model for all

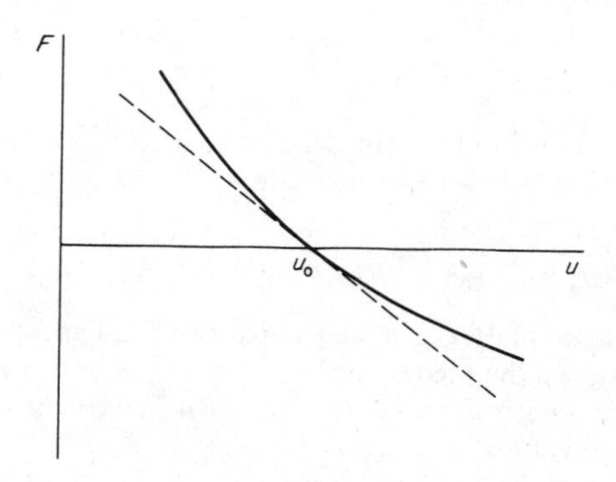

Figure 5.5 Force on the particle in the two potentials of figure 5.4. u_0 is the equilibrium point, at which the force is zero

small vibrations about equilibrium. It is easy to make the argument more rigorous, and at the same time derive an expression for the frequency of small vibrations. We saw that near the equilibrium point $u = u_0$ the potential function essentially coincides with a parabola. We may express the parabola as

$$V^{\mathrm{h}}(u) = V(u_0) + \tfrac{1}{2}B(u - u_0)^2 \tag{5.20}$$

where B is a constant which we shall determine, and the factor $\frac{1}{2}$ is put in for convenience. Equation 5.20 looks like the beginning of a power series in $u - u_0$, so we assume that near enough to u_0 the potential may be written as

$$V(u) = V(u_0) + A(u - u_0) + \tfrac{1}{2}B(u - u_0)^2 + \tfrac{1}{6}C(u - u_0)^3 + \ldots \tag{5.21}$$

This may be familiar as a statement of Taylor's theorem. In order to find the coefficients A, B, C, etc. we proceed as follows. First differentiate both sides with respect to u

$$\frac{\mathrm{d}V}{\mathrm{d}u} = A + B(u - u_0) + \tfrac{1}{2}C(u - u_0)^2 + \ldots \tag{5.22}$$

Next put $u = u_0$, so that all the terms except A on the right-hand side vanish

$$\left(\frac{\mathrm{d}V}{\mathrm{d}u}\right)_{u = u_0} = A \tag{5.23}$$

where the left-hand side is the value of the derivative evaluated at $u = u_0$. In the present case, u_0 is an equilibrium point, and it can be seen from equation 5.19 that $A = 0$. This was assumed in writing down equation 5.20. The procedure for finding B should now be obvious. Differentiating equation 5.22 with respect to u, and putting $u = u_0$, we find

$$\left(\frac{\mathrm{d}^2V}{\mathrm{d}u^2}\right)_{u = u_0} = B \tag{5.24}$$

Similarly the coefficient C is given by

$$\left(\frac{\mathrm{d}^3V}{\mathrm{d}u^3}\right)_{u = u_0} = C \tag{5.25}$$

and similar expressions hold for the further coefficients in the power series of equation 5.21. The power series expansion of $V(u)$ about the equilibrium point u_0 is therefore

$$V(u) = V(u_0) + \tfrac{1}{2}B(u - u_0)^2 + \tfrac{1}{6}C(u - u_0)^3 + \ldots \tag{5.26}$$

with the coefficients B and C given by equations 5.24 and 5.25 respectively.

Near enough to u_0, the higher terms $\frac{1}{6}C(u - u_0)^3$ and so on are generally negligible compared with the second term $\frac{1}{2}B(u - u_0)^2$. In fact the ratio of the third term to the second is

$$\frac{1}{3}\frac{C}{B}(u - u_0) \tag{5.27}$$

which becomes smaller and smaller as u approaches u_0. We conclude, finally, that the parabola of equation 5.20 is indeed of the same form as $V(u)$ for u sufficiently close to u_0. The use made of Taylor's theorem is just the analytic equivalent of the geometrical method summarised in figure 5.4.

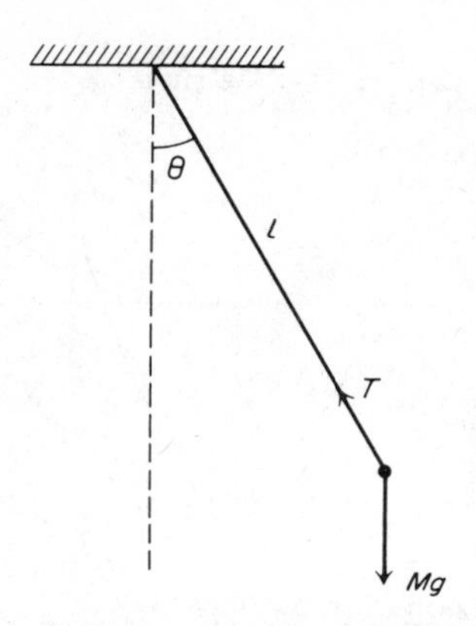

Figure 5.6 Simple pendulum consisting of a mass *M* on a string of length *l*. The forces on the particle are its weight *Mg* and the tension *T* in the string

The argument just given has no pretensions to mathematical rigour. In particular, we did not discuss the convergence of the power series of equation 5.21 or even its existence. These matters are discussed in suitable texts on mathematical analysis. For the potentials we come across in physics, there are usually no difficulties in using Taylor's theorem.

It remains to find the frequency of the small vibrations. Comparing equation 5.20 with the expression for the harmonic oscillator potential, equation 5.8, we see that B is identical to the 'spring constant' K. The angular frequency is therefore given by equation 1.3

$$\omega = (B/m)^{1/2} \tag{5.28}$$

with the value of B in equation 5.24.

Perhaps the most familiar example of small vibrations about an equilibrium point is the simple pendulum, figure 5.6. It will be recalled that the restoring force when the string makes an angle θ with the vertical is proportional to $\sin\theta$. For small vibrations, the approximation $\sin\theta = \theta$ is accurate, so the restoring force is proportional to θ, and the motion is simple harmonic. The argument is rephrased in terms of the potential function in worked example 5.1.

Worked example 5.1

Find the potential function for the simple pendulum of figure 5.6. Expand the potential function in powers of θ for small θ, and hence find the frequency of small vibrations.

Answer

The motion of the particle is a circle about the point of suspension. The tension T is at right angles to this circle, so T does not contribute to the acceleration of the particle. The acceleration is due to the resolved component of the weight, namely, $Mg \sin \theta$, so the equation of motion is

$$Ml \frac{d^2\theta}{dt^2} = -Mg \sin \theta$$

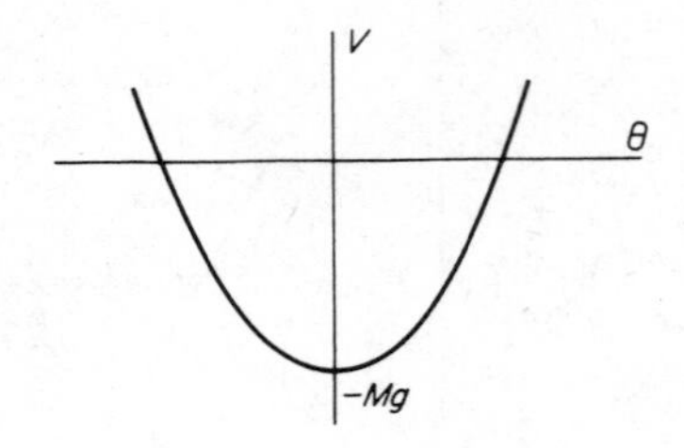

With a force of $-Mg \sin \theta$, equation 5.15 gives

$$V(\theta) = -Mg \cos \theta$$

as shown in the sketch above. The derivatives of V are

$$\frac{dV}{d\theta} = Mg \sin \theta = 0 \qquad \text{at } \theta = 0$$

$$\frac{d^2V}{d\theta^2} = Mg \cos \theta = Mg \qquad \text{at } \theta = 0$$

In the notation of the text, $B = Mg$. In this problem, Ml appears in place of the mass m of equation 5.1, so the angular frequency of small vibrations is given by equation 5.28 with the appropriate substitutions:

$$\omega = (Mg/Ml)^{1/2} = (g/l)^{1/2}$$

which is the usual result.

Instead of the above use of the general method, we could have used the expansion of $\cos \theta$ for small θ to write

$$V(\theta) = -Mg + \tfrac{1}{2}Mg\theta^2$$

whence $B = Mg$ as before.

Worked example 5.2

A particle of mass m moves in the potential

$$V(u) = -\frac{V_0 a^2}{a^2 + u^2}$$

Find the equilibrium point, and the frequency of small vibrations about equilibrium.

Answer

We start by sketching the potential. First, V is always negative. Since V depends only on u^2, we have $V(-u) = V(u)$, so the potential is symmetric about $u = 0$. At $u = 0$, $V = -V_0$. As u increases, the magnitude of V decreases, since the denominator $a^2 + u^2$ increases. Finally, $V \to 0$ as $u \to \infty$ since $a^2 + u^2$ increases indefinitely. The results are summarised in the sketch below.

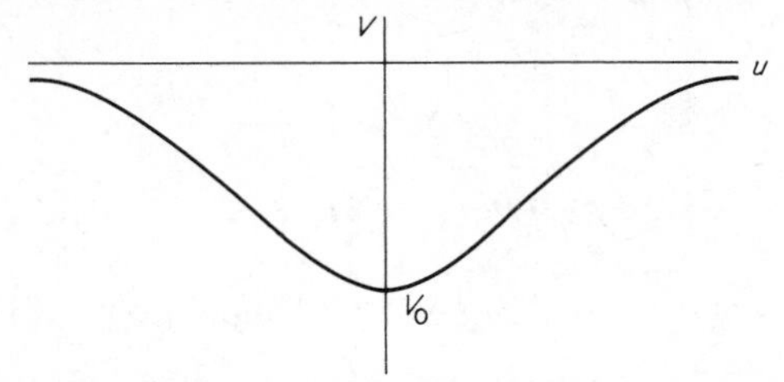

The equilibrium point is obviously $u_0 = 0$, since that is the only minimum in the curve. To find the frequency of small vibrations, we use the general rule.

$$\frac{dV}{du} = \frac{2V_0a^2u}{(a^2 + u^2)^2} = 0 \qquad \text{at } u = 0$$

$$\frac{d^2V}{du^2} = \frac{2V_0a^2}{(a^2 + u^2)^2} - \frac{8V_0a^2u^2}{(a^2 + u^2)^3} = \frac{2V_0}{a^2} \qquad \text{at } u = 0$$

The frequency of small vibrations is given by equation 5.28

$$\omega = (2V_0/ma^2)^{1/2}$$

Note that here, as in worked example 5.1, we can expand the potential by a special method, rather than use the general method. Here we may expand the denominator in $V(u)$ by the binomial theorem

$$\frac{1}{a^2 + u^2} = \frac{1}{a^2}\left(1 - \frac{u^2}{a^2}\right)$$

as far as terms in u^2. This gives

$$V(u) = -V_0 + V_0\frac{u^2}{a^2}$$

By comparison with equation 5.20, we see that $B = 2V_0/a^2$, and the result follows. The advantage of the general method, of course, is that it always works, even though it may be more cumbersome in some particular cases.

5.2 Damped Harmonic Oscillator

We have now seen the significance of the simple harmonic oscillator as a model for small vibrations of any system. In this section, a damping, or dissipative, force is added to the equation of motion of the oscillator. The simplest damping force,

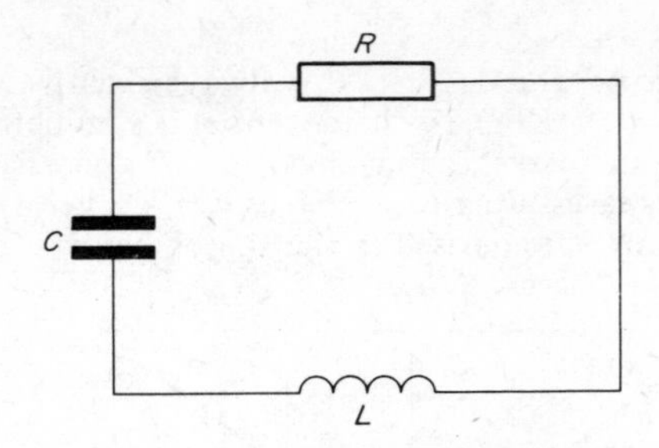

Figure 5.7 *LCR* circuit

and the only one considered here, is one that is simply proportional to the velocity, as in equation 5.16.

$$F = -b \frac{\mathrm{d}u}{\mathrm{d}t} \tag{5.29}$$

The force is written with a minus sign because the damping force is always opposed to the velocity. As is to be expected, the damping force continuously takes energy out of the system, because the rate of working is

$$F \frac{\mathrm{d}u}{\mathrm{d}t} = -b \left(\frac{\mathrm{d}u}{\mathrm{d}t} \right)^2 \tag{5.30}$$

which is always negative. Since energy is taken out by the damping force, the oscillations of the system eventually die away.

With the force of equation 5.29 added, the equation of motion of the oscillator becomes

$$m \frac{\mathrm{d}^2u}{\mathrm{d}t^2} + b \frac{\mathrm{d}u}{\mathrm{d}t} + Ku = 0 \tag{5.31}$$

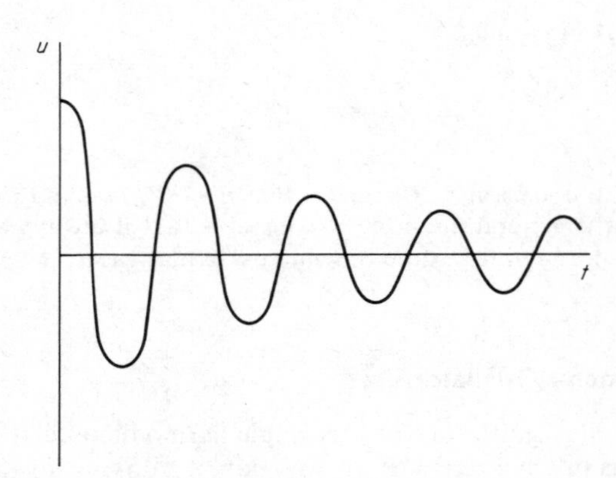

Figure 5.8 Expected form of solution of damped oscillator equation

A well-known example to which this equation of motion applies is the *LCR* circuit shown in figure 5.7. Kirchhoff's law gives for the voltage in the circuit

$$L\frac{\mathrm{d}I}{\mathrm{d}t} + RI + \frac{q}{C} = 0 \tag{5.32}$$

where q is the charge on one plate of the capacitor, and I is the instantaneous current. Since $I = \mathrm{d}q/\mathrm{d}t$, equation 5.32 may be written

$$L\frac{\mathrm{d}^2 q}{\mathrm{d}t^2} + R\frac{\mathrm{d}q}{\mathrm{d}t} + \frac{1}{C}q = 0 \tag{5.33}$$

which has the same form as equation 5.31.

We may expect that the solution of equation 5.31 has something like the form shown in figure 5.8, since damping causes decay of the oscillations. To find the solution, the trial function

$$u = u_0 \exp(\lambda t) \tag{5.34}$$

is substituted. If the damping force is not present, i.e. if $b = 0$, we know that the solution of equation 5.31 is $u_0 \exp(i\omega t)$, which has the form of equation 5.34 with λ taking the pure imaginary value $i\omega$. In the presence of damping, λ may be shown to have a real part as well as an imaginary part. In fact, putting equation 5.34 into equation 5.31 we find

$$(m\lambda^2 + b\lambda + K)u_0 \exp(\lambda t) = 0 \tag{5.35}$$

The trial function is therefore a solution of the equation provided that

$$m\lambda^2 + b\lambda + K = 0 \tag{5.36}$$

which gives

$$\lambda = \frac{-b \pm (b^2 - 4Km)^{1/2}}{2m} \tag{5.37}$$

The form of the solution depends on the sign of $b^2 - 4Km$, since for $b^2 < 4Km$ the square root is imaginary, and the two values of λ are complex, whereas for $b^2 > 4Km$ the square root is real, and both values of λ are real. We take the two cases in turn.

First, with $b^2 < 4Km$, that is with *light damping*, we have

$$\lambda = -\frac{b}{2m} \pm i\omega_1 \tag{5.38}$$

where

$$\omega_1{}^2 = \frac{4Km - b^2}{4m^2} = \omega_0{}^2 - \frac{b^2}{4m^2} \tag{5.39}$$

We have put

$$\omega_0{}^2 = K/m \tag{5.40}$$

so that ω_0 is the frequency of the undamped oscillator. It is convenient to define

$$\Gamma = b/m \tag{5.41}$$

so that the solutions for light damping may be written

$$u = u_0 \exp(-\tfrac{1}{2}\Gamma t) \exp(\pm i\omega_1 t) \tag{5.42}$$

The meaning of the ± sign needs explanation. Like the equation of the undamped oscillator, equation 5.31 is second order in t, and we may therefore expect two independent solutions, like $\sin \omega_0 t$ and $\cos \omega_0 t$ in the undamped oscillator. The trial function, equation 5.34, is written with the usual convention that u_0 is complex and the real part of the product is taken. Since either sign of ω_1 gives a function which satisfies the differential equation, the general solution is a linear combination

$$u = u_1 \exp(-\tfrac{1}{2}\Gamma t) \exp(i\omega_1 t) + u_2 \exp(-\tfrac{1}{2}\Gamma t) \exp(-i\omega_1 t) \tag{5.43}$$

where both u_1 and u_2 are complex, and the real part is taken. This obviously amounts to taking a linear combination, with real coefficients, of two solutions, one with $\sin \omega_1 t$ and one with $\cos \omega_1 t$

$$u = u_0 \exp(-\tfrac{1}{2}\Gamma t) \sin \omega_1 t \tag{5.44}$$

$$u = u_0' \exp(-\tfrac{1}{2}\Gamma t) \cos \omega_1 t \tag{5.45}$$

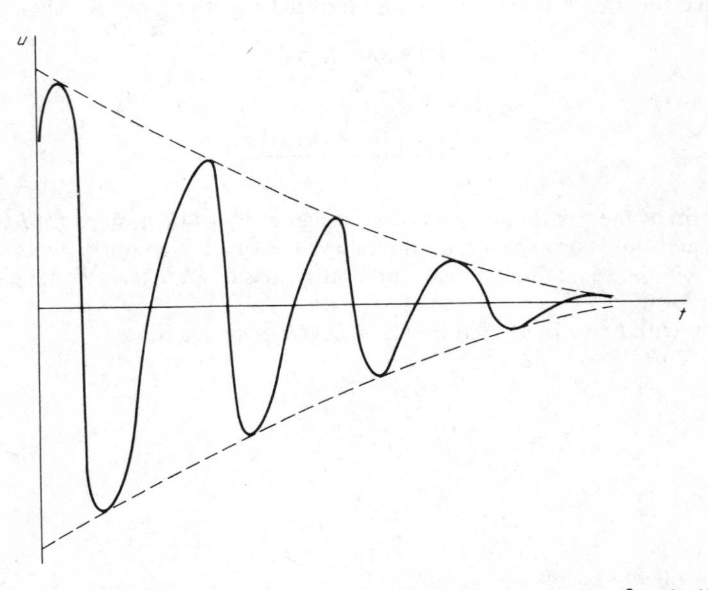

Figure 5.9 Solution of the damped oscillator equation for light damping $b^2 < 4mK$. The dashed curve is the modulation function $\pm u_0 \exp(-\frac{1}{2}\Gamma t)$

Equation 5.42 is therefore a shorthand equivalent of equations 5.44 and 5.45. The coefficients of the sine and cosine terms are determined by the initial conditions. The method is shown in worked example 5.3.

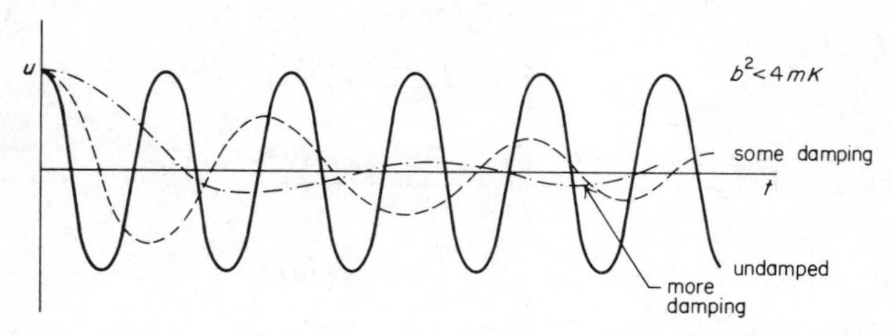

Figure 5.10 Solutions of the damped oscillator equation for various values of the damping constant *b*

It is instructive to sketch the solution, equation 5.42. The method is similar to that used for the beat pattern of figure 1.10. The term $u_0 \exp(-\frac{1}{2}\Gamma t)$ is regarded as an envelope function $u_{\text{MOD}}(t)$, and the sine or cosine term in equation 5.44 or 5.45 describes oscillations between the limits $\pm |u_{\text{MOD}}(t)|$. The solution is therefore as sketched in figure 5.9 and as anticipated it consists of decaying oscillations. Equations 5.38 and 5.41 show how the solution varies as the damping constant b increases. For $b = 0$, we have the usual undamped oscillations of angular frequency ω_0. As b increases from zero, the oscillations decay more and more rapidly (equation 5.41), and their frequency decreases (equation 5.39). These results are summarised in figure 5.10.

The exponentially decaying amplitude function $u_{\text{MOD}}(t) = \exp(-\frac{1}{2}\Gamma t)$ is precisely analogous to the spatially decaying amplitude $u_{\text{MOD}}(x) = \exp(-\alpha x)$ which we came across in our discussion of spatial attenuation of waves in section 2.6. Consequently, the main features of that discussion can be taken over. First, $u_{\text{MOD}}(t)$ can be plotted against t on log–linear graph paper, to find a straight line. Secondly, the decay of the amplitude can be expressed as a certain number of dB per second. The methods are precisely the same as were described in detail in section 2.6.

At the particular value of the damping constant $b^2 = 4Km$, which is called *critical damping*, it is seen from equation 5.37 that the two values of λ coincide at

$$\lambda = -b/2m \tag{5.46}$$

One solution of the equation is therefore

$$u = u_0 \exp(-\tfrac{1}{2}\Gamma t) \tag{5.47}$$

with $\Gamma = b/m$ as before. This solution is of course the limiting case of the light damping solution with $\omega_1 = 0$. Since we are solving equation 5.31, which is of second order, there must be a second solution, which is in fact

$$u = u_0 t \exp(-\tfrac{1}{2}\Gamma t) \tag{5.48}$$

The proof that this is a solution of equation 5.31 if $b^2 = 4Km$ is left as a problem.

With *heavy damping,* $b^2 > 4Km$, equation 5.37 gives two real values for λ, both of them negative

$$\lambda = \lambda_1 \text{ or } \lambda_2 \tag{5.49}$$

$$\lambda_1 = -\frac{b}{2m} - \frac{(b^2 - 4Km)^{1/2}}{2m} \tag{5.50}$$

$$\lambda_2 = -\frac{b}{2m} + \frac{(b^2 - 4Km)^{1/2}}{2m} \tag{5.51}$$

The general solution is a sum of two decaying exponentials:

$$u = u_1 \exp(\lambda_1 t) + u_2 \exp(\lambda_2 t) \tag{5.52}$$

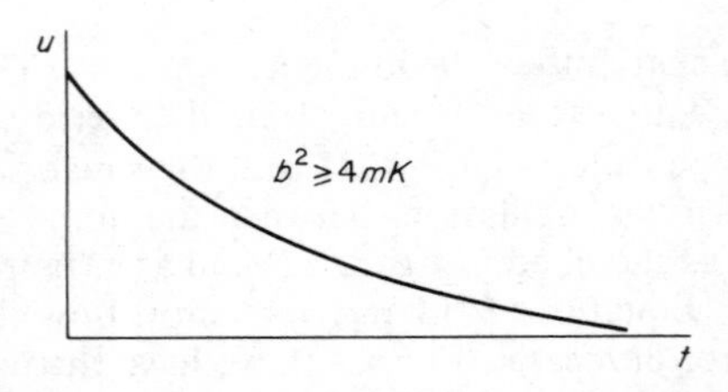

Figure 5.11 Sketch of solution for critical or heavy damping

For both critical and heavy damping, the solution is simply a decaying function with no oscillations, as sketched in figure 5.11.

Worked example 5.3

A lightly damped oscillator has $\omega_1 = 10^7 \text{ s}^{-1}$ and $\Gamma = 10^6 \text{ s}^{-1}$. It is set into motion with initial displacement $u = 10^{-1}$ m and initial velocity $du/dt = 10^5 \text{ m s}^{-1}$. Find the subsequent motion.

Answer

The motion must be a linear combination of the two solutions given by equations 5.44 and 5.45

$$u = u_1 \exp(-\tfrac{1}{2}\Gamma t) \sin \omega_1 t + u_2 \exp(-\tfrac{1}{2}\Gamma t) \cos \omega_1 t$$

The velocity is

$$\frac{du}{dt} = \omega_1 u_1 \exp(-\tfrac{1}{2}\Gamma t) \cos \omega_1 t - \omega_1 u_2 \exp(-\tfrac{1}{2}\Gamma t) \sin \omega_1 t$$
$$-\tfrac{1}{2}\Gamma u_1 \exp(-\tfrac{1}{2}\Gamma t) \sin \omega_1 t - \tfrac{1}{2}\Gamma u_2 \exp(-\tfrac{1}{2}\Gamma t) \cos \omega_1 t$$

The initial values ($t = 0$) are

$$u = u_2$$

$$\frac{du}{dt} = \omega_1 u_1 - \tfrac{1}{2}\Gamma u_2$$

The first equation gives $u_2 = 10^{-1}$ m. Substituting the initial value of the velocity into the second equation, we find

$$10^5 = 10^7 u_1 - 5 \times 10^4$$

so that

$$u_1 = 5 \times 10^{-3} \text{ m}$$

The motion is therefore

$$u = 5 \times 10^{-3} \exp(-\tfrac{1}{2}\Gamma t) \sin \omega_1 t + 10^{-1} \exp(-\tfrac{1}{2}\Gamma t) \cos \omega_1 t$$

Worked example 5.4

Show that the limiting value as $\lambda_1 \to \lambda_2$ of equation 5.52 may be written

$$u = u_3 \exp(\lambda_1 t) + u_4 t \exp(\lambda_1 t)$$

and find expressions for u_3 and u_4.

Answer

Equation 5.52 may be rewritten as

$$u = \tfrac{1}{2}(u_1 + u_2)[\exp(\lambda_1 t) + \exp(\lambda_2 t)] + \tfrac{1}{2}(u_1 - u_2)[\exp(\lambda_1 t) - \exp(\lambda_2 t)]$$

As $\lambda_1 \to \lambda_2$, the first term in square brackets has the simple limit

$$\exp(\lambda_1 t) + \exp(\lambda_2 t) \to 2 \exp(\lambda_1 t)$$

The second term may be written

$$\exp(\lambda_1 t) - \exp(\lambda_2 t) = (\lambda_1 - \lambda_2) \frac{d}{d\lambda} \exp(\lambda_1 t)$$

$$= (\lambda_1 - \lambda_2) t \exp(\lambda_1 t)$$

Therefore the limiting form is

$$u = (u_1 + u_2) \exp(\lambda_1 t) + \tfrac{1}{2}(u_1 - u_2)(\lambda_1 - \lambda_2) t \exp(\lambda_1 t)$$

which is of the form quoted in the question, with

$$u_3 = u_1 + u_2$$

$$u_4 = \tfrac{1}{2}(u_1 - u_2)(\lambda_1 - \lambda_2)$$

Comment

This example shows how the second solution for critical damping, quoted without proof in equation 5.48, arises as a limiting form from heavy damping, and in fact this example constitutes a proof that equation 5.48 is a solution for $b^2 = 4Km$. The proof by direct substitution is left for problem 5.7.

5.3 Resonance

We now come to the main topic of this chapter, namely the response of a damped harmonic oscillator to a harmonic driving force. The main quantities to calculate are the amplitude of the oscillations produced, their phase lag relative to the driving force, and the power absorbed from the driving force by the oscillator. The absorption of power is ultimately by the damping force, and it can be shown that the average power taken from the driving force is equal to the average power absorbed by damping.

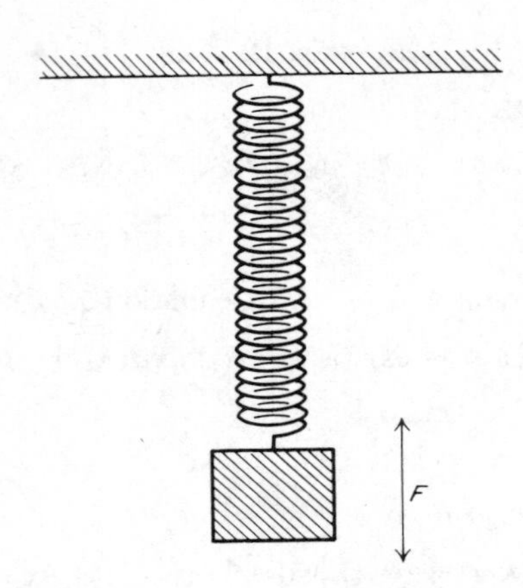

Figure 5.12 Driven mechanical oscillator

The discussion will apply to two rather distinct types of system. First, there are rather simple systems to which the equation to be used applies obviously and directly. Examples of this type are; a mechanical oscillator with friction subject to a periodic driving force (figure 5.12) and an *LCR* circuit across which an a.c. voltage is applied (figure 5.13). Second, and more important, we may have an oscillation about equilibrium of a complex system stimulated by an incident wave. As shown in chapter 4, microscopic particles like electrons and neutrons can propagate as waves, so the discussion will also apply to excitation of oscillations about equilibrium by particle beams. The scope of the discussion is therefore very broad, and it will be seen in section 5.5 that the characteristic resonance curve for power absorption is indeed found in a very wide range of physical systems.

The equation of motion of a mechanical oscillator subject to a periodic driving force, figure 5.12, is

$$m\frac{\mathrm{d}^2u}{\mathrm{d}t^2}+b\frac{\mathrm{d}u}{\mathrm{d}t}+Ku=F\cos\omega t \tag{5.53}$$

The choice of phase of the driving force, $\delta=0$ in the general expression $F\cos(\omega t+\delta)$, is not a limitation, since it is equivalent to choosing the origin for time measurement such that $t=0$ corresponds to a point of maximum amplitude F of the driving force.

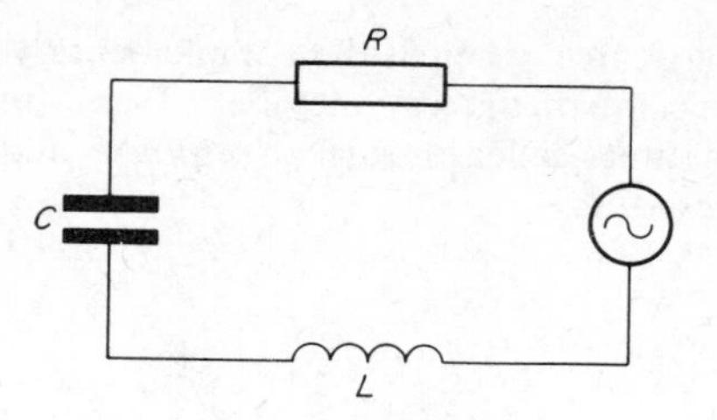

Figure 5.13 Driven *LCR* circuit

Of course the displacement u will be allowed to have a general phase, because the phase of u relative to the driving force is very important.

The equation for the driven *LCR* circuit (figure 5.13) is

$$L\frac{d^2q}{dt^2}+R\frac{dq}{dt}+\frac{1}{C}q=E\cos\omega t \tag{5.54}$$

which is equivalent to equation 5.53. For consistency with our earlier work, the 'mechanical' notation of equation 5.53 will be used.

We start by proving a general result about the solutions of equation 5.53. The equation is of second order, so the general solution involves two arbitrary constants, which are determined in the usual way by the initial conditions. The left-hand side of the equation is also linear in u, and the linearity means that the following theorem can be proved. The *general* solution of equation 5.53 can be written as

$$u(t)=u_1(t)+u_2(t) \tag{5.55}$$

where $u_1(t)$ is the *general* solution of the *homogeneous* equation

$$m\frac{d^2u_1}{dt^2}+b\frac{du_1}{dt}+Ku_1=0 \tag{5.56}$$

and $u_2(t)$ is any *particular* solution of equation 5.53. The proof is simple, and similar to that of the superposition principle in equation 1.19. Since $u(t)$ and $u_2(t)$ are both solutions of equation 5.53, their difference, which is $u_1(t)$, satisfies

$$\begin{aligned}m\frac{d^2}{dt^2}(u-u_2)+b\frac{d}{dt}(u-u_2)+K(u-u_2)&=F\cos\omega t-F\cos\omega t\\&=0\end{aligned} \tag{5.57}$$

which is equation 5.56. This completes the proof of the theorem.

The point of the theorem just proved is that the two arbitrary constants in the general solution $u(t)$ of equation 5.53 appear in $u_1(t)$, which satisfies the simpler equation 5.56. This is the equation of the damped oscillator, which we discussed in detail in section 5.2. Our present task is, therefore, to find just a particular solution of equation 5.53. The function $u_2(t)$ is simply called the *particular integral*, and $u_1(t)$ the *complementary function*. As sketched in figure 5.14, the particular solution may be expected to continue oscillating indefinitely at the frequency ω of the driving force, whereas the complementary function decays exponentially in time. Because of the way they behave as a function of time, the particular integral is sometimes called the *driven response*, and the complementary function, the *transient response.*

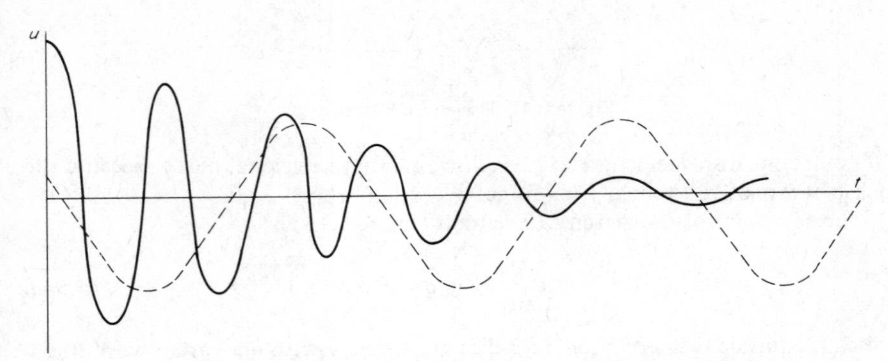

Figure 5.14 Particular integral (dashed line) and complementary function (solid line)

We now concentrate on the particular integral, or driven response. Rewrite equation 5.53 as

$$m\frac{\mathrm{d}^2u}{\mathrm{d}t^2}+b\frac{\mathrm{d}u}{\mathrm{d}t}+Ku=F\exp(-i\omega t) \tag{5.58}$$

where as usual it is understood that we are to take the real part of all expressions. We assume that the response is at the same frequency as the driving force, so we substitute the trial solution

$$u=u_0\exp(-i\omega t) \tag{5.59}$$

to find

$$(-m\omega^2-i\omega b+K)u_0=F \tag{5.60}$$

or

$$u_0=\frac{F}{m(\omega_0^2-\omega^2-i\omega\Gamma)} \tag{5.61}$$

where equations 5.40 and 5.41 are used to introduce the frequency of the undamped oscillator, ω_0, and the damping constant Γ. Equation 5.61 gives a complex value for u_0, which shows that as expected there is a phase difference between the driving force and the response. It is convenient to separate the real and imaginary parts of u_0 by rationalising the denominator

$$u_0=\frac{F}{m}(A+iB) \tag{5.62}$$

with

$$A = \frac{\omega_0^2 - \omega^2}{(\omega_0^2 - \omega^2)^2 + \omega^2\Gamma^2} \tag{5.63}$$

$$B = \frac{\omega\Gamma}{(\omega_0^2 - \omega^2)^2 + \omega^2\Gamma^2} \tag{5.64}$$

Taking the real part of equation 5.59 explicitly

$$u = \frac{F}{m} \operatorname{Re}(A + iB)(\cos \omega t - i \sin \omega t) \tag{5.65}$$

$$= \frac{F}{m}(A \cos \omega t + B \sin \omega t) \tag{5.66}$$

Since the first term of equation 5.66 is in phase with the driving force, A is sometimes called the *in-phase amplitude*, and B the *out-of-phase amplitude.* It is instructive to sketch A and B as functions of the driving frequency ω. The behaviour of both is governed by the *resonant denominator* D

$$D = (\omega_0^2 - \omega^2)^2 + \omega^2\Gamma^2 \tag{5.67}$$

D has a minimum at $\omega = \omega_0$, and if the damping constant Γ is small, the minimum is very sharp. Equation 5.64 then shows that B has a maximum near ω_0, tends to zero as $\omega \to 0$ and as $\omega \to \infty$, and is always positive. A is equal to $1/\omega_0^2$ at $\omega = 0$, zero at $\omega = \omega_0$, and tends to zero from negative values as $\omega \to \infty$. These results are shown in figures 5.15 and 5.16.

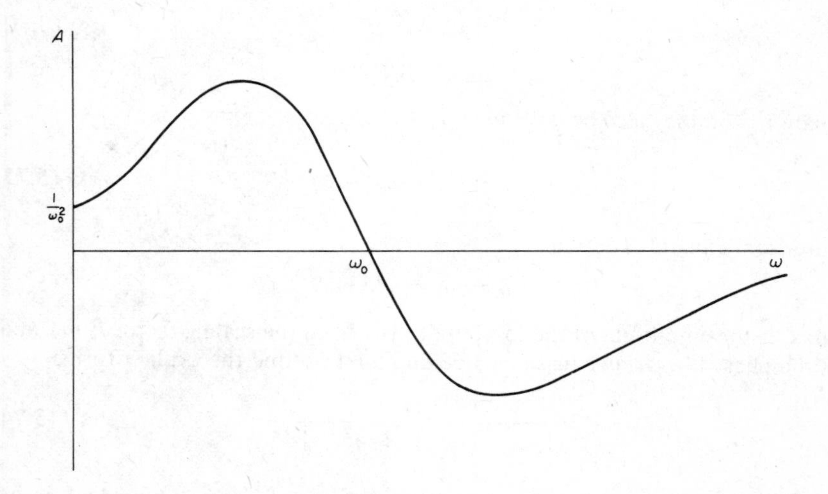

Figure 5.15 In-phase amplitude A as a function of driving frequency

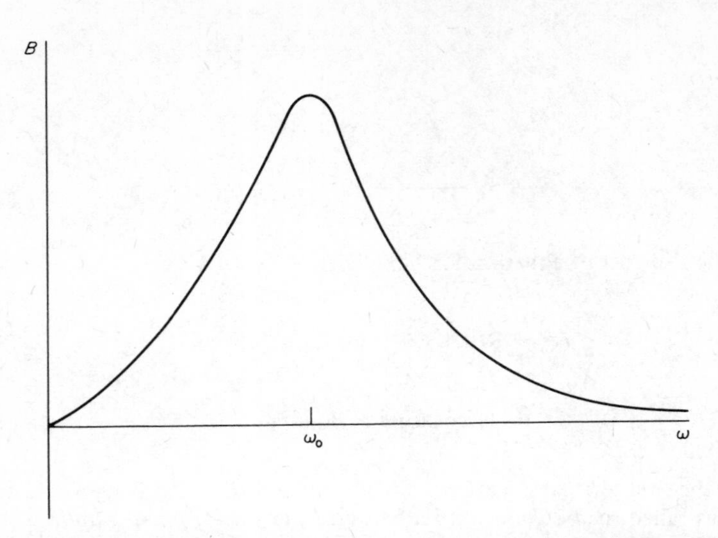

Figure 5.16 Out-of-phase amplitude B as a function of driving frequency

We may also look at the overall amplitude of the response and its phase lag relative to the driving force. We write equation 5.66 as

$$u = \frac{F}{m} C(\cos \delta \cos \omega t + \sin \delta \sin \omega t) \tag{5.68}$$

where

$$C = (A^2 + B^2)^{1/2} \tag{5.69}$$

$$\cos \delta = A/C \tag{5.70}$$

$$\sin \delta = B/C \tag{5.71}$$

Equation 5.68 may also be written

$$u = \frac{F}{m} C \cos(\omega t - \delta) \tag{5.72}$$

where from equations 5.70 and 5.71

$$\delta = \tan^{-1}(B/A) \tag{5.73}$$

Thus C is the amplitude of the response (apart from the scaling factor F/m), and δ is the phase lag. Using equations 5.63 and 5.64 we find the explicit forms

$$C = \frac{1}{[(\omega_0^2 - \omega^2)^2 + \omega^2 \Gamma^2]^{1/2}} \tag{5.74}$$

$$\delta = \tan^{-1} \frac{\omega \Gamma}{\omega_0^2 - \omega^2} \tag{5.75}$$

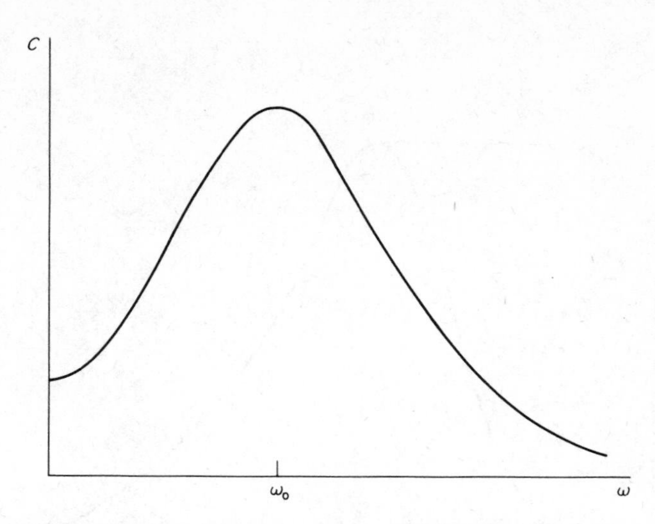

Figure 5.17 Total amplitude C as a function of driving frequency

The amplitude C is equal to $1/\omega_0{}^2$ at $\omega = 0$, has a maximum at $\omega = \omega_0$, and tends to zero as $\omega \to \infty$. The phase lag δ is 0 at $\omega = 0, \pi/2$ (that is, $\tan^{-1} \infty$) at $\omega = \omega_0$, and tends to π ($\tan^{-1}$ (0)) as $\omega \to \infty$. These results are shown in figures 5.17 and 5.18. The phase lag is of course zero at zero frequency and increases with frequency as the system lags progressively further behind the driving force. As might be expected, the amplitude of the response is a maximum at $\omega = \omega_0$.

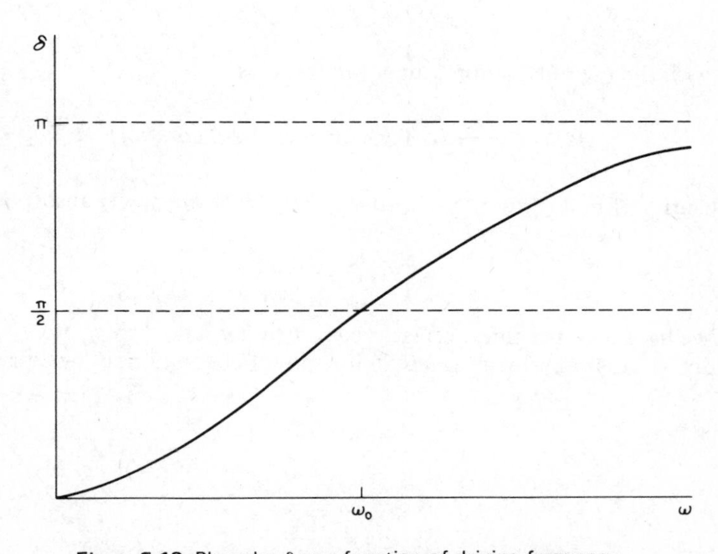

Figure 5.18 Phase lag δ as a function of driving frequency

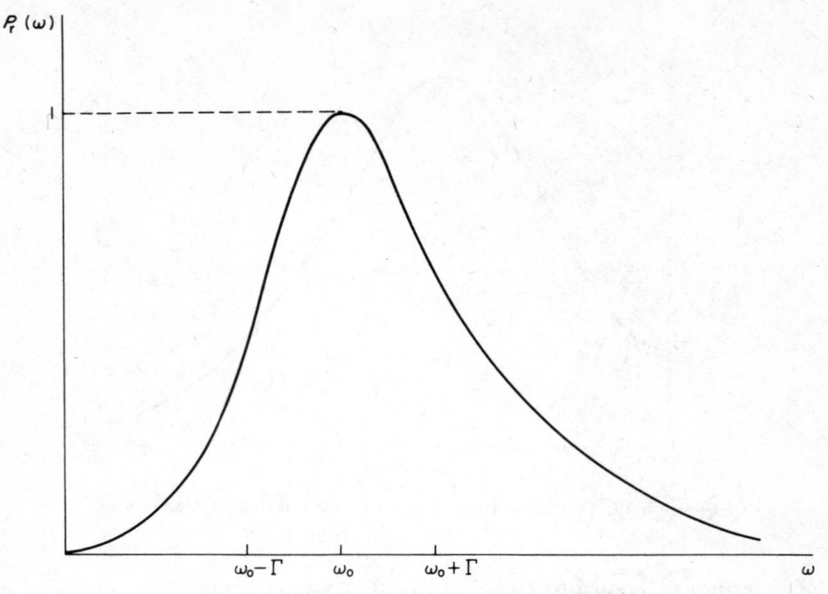

Figure 5.19 Power absorption function $P_r(\omega)$

We can now calculate the rate of power absorption out of the driving force, which of course is the quantity of most direct physical interest. The instantaneous rate of working is

$$P(t) = F \cos \omega t \frac{\mathrm{d}u}{\mathrm{d}t} \tag{5.76}$$

which with the use of equation 5.66 for u becomes

$$P(t) = \frac{F^2\omega}{m} (-A \cos \omega t \sin \omega t + B \cos^2 \omega t) \tag{5.77}$$

The quantity of most interest is the *time average* P of the power absorption

$$P = \langle P(t) \rangle = \frac{F^2 \omega B}{2m} \tag{5.78}$$

where we have used the time averages $\langle \cos \omega t \sin \omega t \rangle = 0, \langle \cos^2 \omega t \rangle = \frac{1}{2}$ (see appendix 1). It is convenient to rewrite the power absorption, using equation 5.64 for B, as

$$P = P_0 P_r(\omega) \tag{5.79}$$

with

$$P_0 = \frac{F^2}{2m\Gamma} \tag{5.80}$$

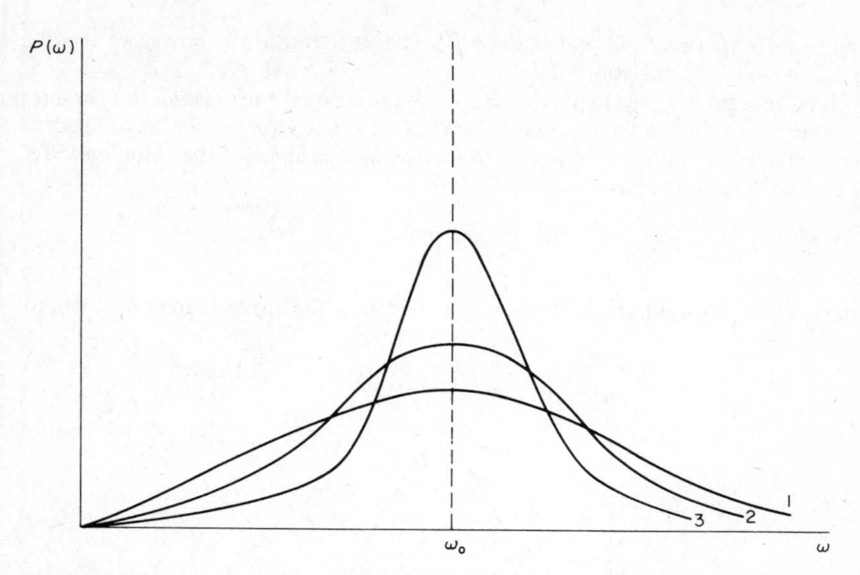

Figure 5.20 Resonance curves for three different values of Γ: $\Gamma_1 > \Gamma_2 > \Gamma_3$ where the subscripts correspond to the labelling of the curves

and

$$P_r(\omega) = \frac{\omega^2\Gamma^2}{(\omega_0^2 - \omega^2)^2 + \omega^2\Gamma^2} \tag{5.81}$$

The *resonance function* $P_r(\omega)$, which is sketched in figure 5.19, has a maximum value of unity at $\omega = \omega_0$.

It can be seen from equation 5.78 that only the out-of-phase amplitude B contributes to the net power absorption. The reason is apparent from equation 5.76: only the part of the velocity $\mathrm{d}u/\mathrm{d}t$ which is in phase with the driving force contributes to $\langle P(t) \rangle$ and the term $-\omega A \sin \omega t$ in the velocity is $\pi/2$ out of phase with the driving force. Because of this difference between the two terms in the response, A is sometimes called the *elastic amplitude,* and B the *absorptive amplitude.* We can now see more significance in figures 5.15 and 5.16 for A and B. In particular, at the *resonant frequency*, $\omega = \omega_0$, the elastic amplitude is zero and the absorptive amplitude is large. Equivalently, the phase lag is exactly $\pi/2$ at the resonant frequency.

It is very important to see how the form of the power absorption curve depends on the damping constant Γ. First, from equation 5.80 it is seen that the peak value of power absorption increases as Γ decreases. Second, equation 5.81 shows that the *width* of the resonance curve $P_r(\omega)$ is proportional to Γ. That is, as indicated on figure 5.19, the curve dies away fairly rapidly on either side of $\omega = \omega_0 \pm \Gamma$. Thus as Γ decreases, the resonance curve becomes higher and narrower and the resonance becomes sharper. Small Γ corresponds to small damping of the free oscillator (equation 5.41), and free oscillations which are sustained for a relatively long time, (equation 5.42 and figure 5.9). The variation of the power absorption with Γ is sketched in figure 5.20.

It is sometimes convenient to use a parameter called the *Q-value* to characterise the sharpness of a resonance curve. Q stands for quality factor, and is defined so that increasing Q-value corresponds to increasing sharpness of the resonance. To be precise Q is defined as

$$Q = \frac{\omega_0}{\Delta\omega} \tag{5.82}$$

where, as shown in figure 5.21, $\Delta\omega$ is the width of the curve at the point where

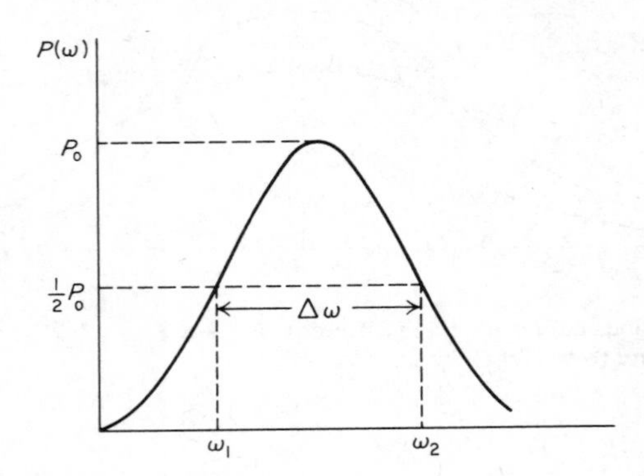

Figure 5.21 $\Delta\omega$ is width at half-power level

the power absorption is half the maximum value. Obviously $\Delta\omega = \omega_2 - \omega_1$, where ω_2 and ω_1 are the two solutions of

$$P_r(\omega) = \tfrac{1}{2} \tag{5.83}$$

and $P_r(\omega)$ is the resonance function of equation 5.81. The solution of equation 5.83 is straightforward, and is left for problem 5.12; the result is

$$\Delta\omega = \omega_2 - \omega_1 = \Gamma \tag{5.84}$$

We therefore have for the Q-value, equation 5.82

$$Q = \omega_0/\Gamma \tag{5.85}$$

That is, the Q-value increases as Γ decreases, in line with the sketches of figure 5.20.

We often deal with resonances in which the decay constant Γ is small compared with the resonance frequency ω_0. As already remarked, this is the case when the oscillations of the undriven system are sustained for a relatively long time. The resonance curve of figure 5.19 is then sharply peaked at ω_0, and we may simplify the resonant denominator D by putting $\omega = \omega_0$ wherever possible

$$\begin{aligned} D &= (\omega_0 + \omega)^2(\omega_0 - \omega)^2 + \omega^2\Gamma^2 \\ &\approx 4\omega_0^2(\omega_0 - \omega)^2 + \omega_0^2\Gamma^2 \end{aligned} \tag{5.86}$$

With this approximation, $P_r(\omega)$ becomes

$$P_L(\omega) = \frac{\frac{1}{4}\Gamma^2}{(\omega_0 - \omega)^2 + \frac{1}{4}\Gamma^2} \tag{5.87}$$

This curve, known as the *lorentzian line shape*, is sketched in figure 5.22. It is symmetrical about $\omega = \omega_0$, and takes its maximum value of unity at $\omega = \omega_0$.

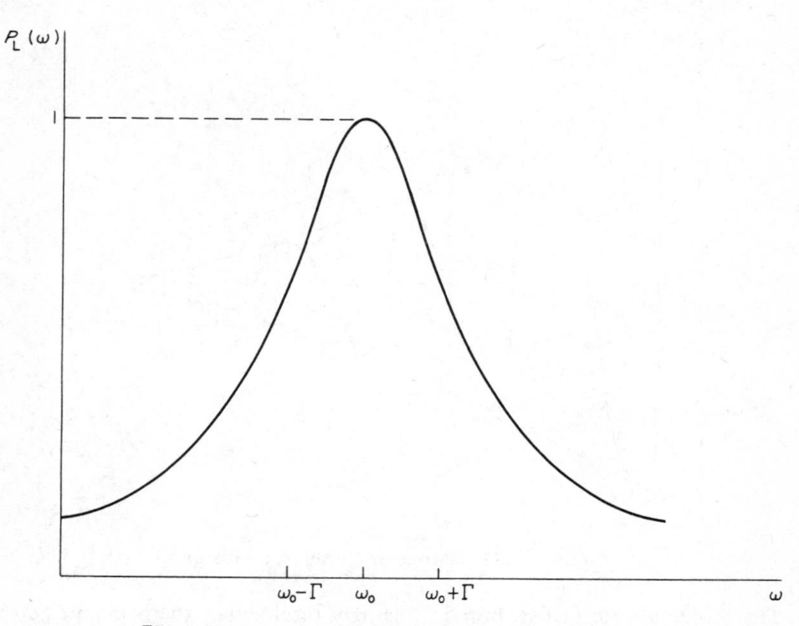

Figure 5.22 The lorentzian line shape of equation 5.87

5.4 Decay Time and Relation to Uncertainty Product

It was shown in section 1.4 that a waveform of duration δt contains a spread of frequencies $\delta\omega$ given by the uncertainty product

$$\delta\omega\,\delta t \approx 2\pi \tag{5.88}$$

If we write the waveform of the decaying oscillator, equation 5.42, as

$$u(t) = u_0 \exp(-t/2\tau) \exp(\pm i\omega_1 t) \tag{5.89}$$

with the *decay time* or *lifetime* τ given by

$$\tau = 1/\Gamma \tag{5.90}$$

we see that the waveform has a duration $\delta t \approx \tau$ as shown in figure 5.23. The waveform therefore contains a frequency spread $\delta\omega$ given by

$$\delta\omega \approx 2\pi/\tau = 2\pi\Gamma \qquad (5.91)$$

Equation 5.89 was quoted in section 1.4, and figure 1.23(b) shows the frequency waveform $|\tilde{u}(\omega)|$.

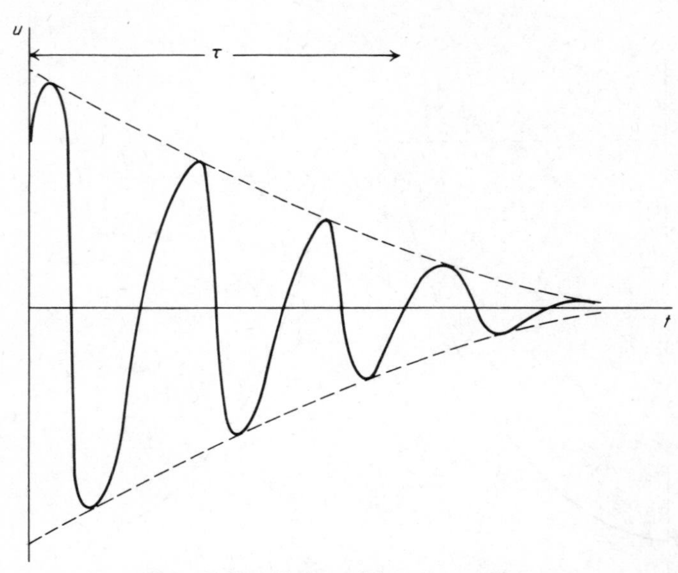

Figure 5.23 Lifetime of decaying oscillator

The principal result of section 5.3, namely the form of the resonance curve, figure 5.19 or 5.22, is now easy to understand. The decaying oscillator contains a range of frequencies $\delta\omega$. If the system is driven with a force $F\cos\omega t$ whose frequency falls within this range, there is a response of some magnitude, and the system removes a significant amount of power from the driving force. Hence the width in frequency of the resonance curve is exactly the same as the frequency spread $\delta\omega$ of the decaying oscillator waveform. And indeed the 'uncertainty principle' estimate of the width, equation 5.91, agrees with equation 5.84 apart from the numerical factor 2π. This numerical factor is of no real significance. As pointed out in section 1.4 the sign $\approx$ is used in equation 5.88 partly because of the latitude in defining $\delta\omega$; equally in finding $\Delta\omega$ the width was taken at power $\frac{1}{2}P_0$, but we could have chosen any power level between 0 and P_0.

It is useful to express the Q-value in terms of the decay time τ. With the use of equations 5.85 and 5.90, we find simply

$$Q = \omega_0\tau \qquad (5.92)$$

Thus the Q-value is a direct measure of the decay time of the free oscillator.

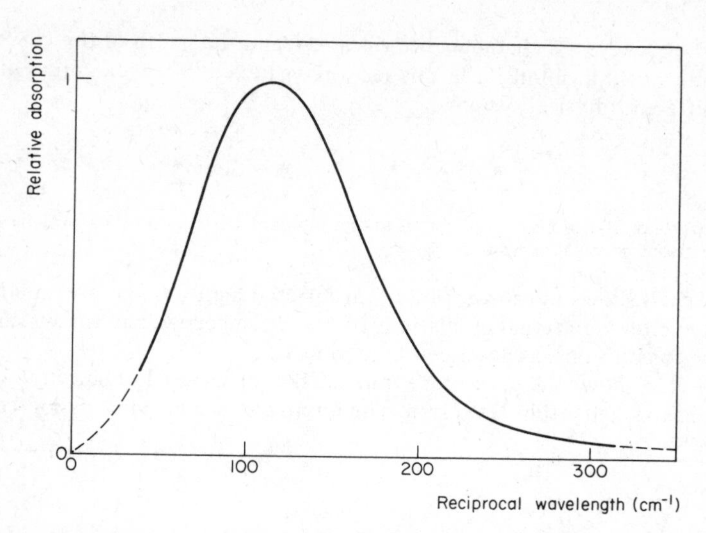

Figure 5.24 Absorption of far infra-red light by high-pressure N_2 gas. The horizontal axis is in units of $K = 1/\lambda$. (Reproduced, with permission, from M. F. Kimmit, *Far Infrared Techniques,* Pion (1970))

5.5 Examples of Resonance

It was shown in section 5.1 that small vibrations about equilibrium are generally simple harmonic oscillations. Therefore, if a system is driven by a periodic force, like that produced by a wave, and the power absorption curve has the resonance form of figure 5.19 or figure 5.22, then the position of the maximum in the curve

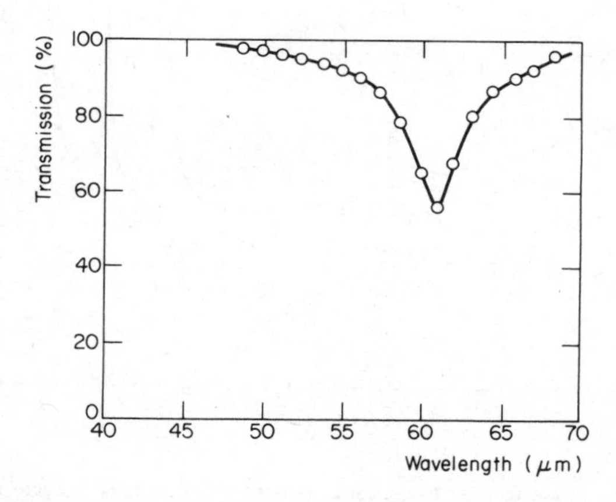

Figure 5.25 Transmission of infra-red radiation at normal incidence through a thin (0.17 μm) sodium chloride film. (Reproduced, with permission, from C. Kittel, *Introduction to Solid State Physics,* 3rd edn, Wiley, New York (1966))

gives the frequency of a characteristic vibration, and the width of the curve gives the lifetime τ of the vibration. In this section we look at a few examples of resonance in different physical systems.

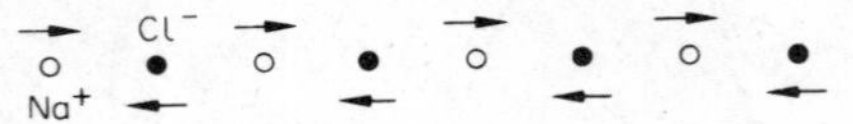

Figure 5.26 Velocities of Na^+ and Cl^- ions at one moment during the long-wavelength optical vibrations of a NaCl crystal

Figure 5.24 shows the absorption of far infrared light by high pressure N_2 gas. The light excites vibrational oscillations of the N_2 molecule, and the width of the line is due to collisions, as described in section 1.5.

Figure 5.25 shows the percentage transmission of infrared radiation at various frequencies through a thin NaCl film. The lorentzian dip in transmission corre-

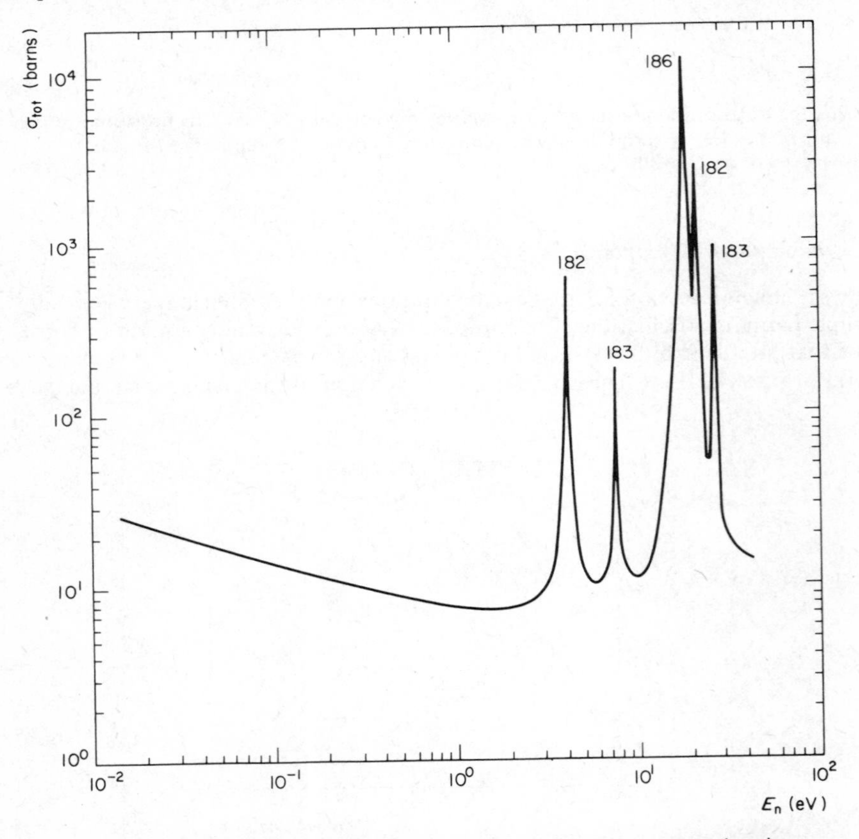

Figure 5.27 The cross section for low-energy neutrons on natural tungsten showing resonances in the isotopes ^{182}W, ^{183}W and ^{186}W. Note the logarithmic scales on both axes. (Reproduced, with permission, from D. F. Jackson, *Nuclear Reactions,* Methuen, London (1970))

sponds to absorption of light by the so-called long wavelength optical vibrations of NaCl, in which the Na^+ and Cl^- ions oscillate in anti-phase, as shown schematically in figure 5.26.

An example of resonances occurring in nuclear physics is shown in figure 5.27. This concerns the capture of low-energy neutrons by a tungsten target. The quantity shown is the cross-section σ for capture of a neutron from the incident beam, which is defined as the probability of capture if the incident beam carries one particle per second and the target contains one nucleus per unit area. It can be seen from this definition that σ has the dimensions of area, and it is measured in units of *barns*

$$1 \text{ barn} = 10^{-28} \text{ m}^2 \tag{5.93}$$

The target for the experimental results shown in figure 5.27 consisted of natural tungsten, which means that it contained the naturally occurring range of tungsten isotopes. Figure 5.27 shows that the cross-section for neutron capture, regarded as a function of neutron energy, consists of a series of sharp resonances. The explanation is that when the incident neutron has one of the resonant values, the neutron is absorbed by the target nucleus, say ^{N}W, to form a compound nucleus, which is an unstable excited state of the nucleus ${}^{N+1}$W. The compound nucleus has a lifetime τ, and decays to the ground state of ${}^{N+1}$W with the emission of a Υ-ray. Thus the compound nucleus may be regarded as a highly unstable radioactive Υ-emitter. The widths of the resonances in figure 5.27 give an estimate of the lifetimes of the compound nuclei. We may translate the energy width δE into a frequency width

$$\delta E = h \delta f \tag{5.94}$$

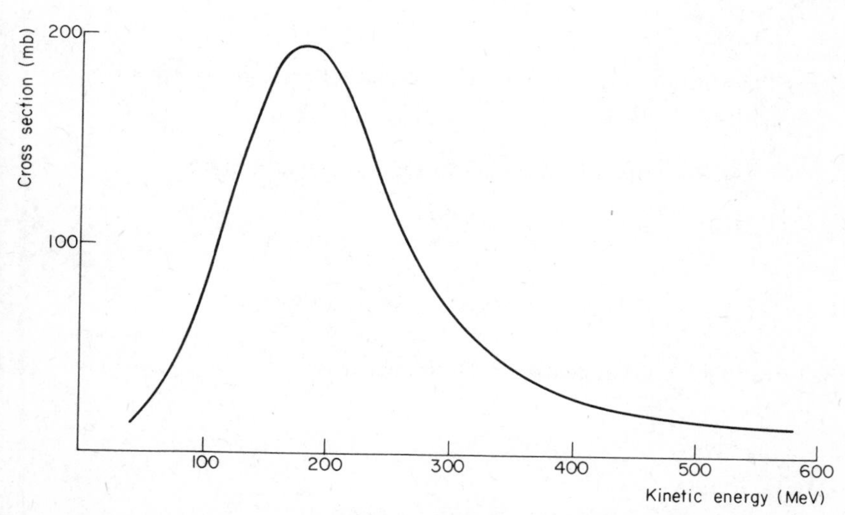

Figure 5.28 Cross-section for the scattering of π^+ mesons on protons as a function of meson energy. (Reproduced, with permission, from I. S. Hughes, *Elementary Particles,* Penguin, Harmondsworth (1972))

and therefore find for the lifetime

$$\tau \approx \frac{1}{\delta f} \approx \frac{h}{\delta E} \tag{5.95}$$

We leave it for problem 5.16 to show that for the resonances of figure 5.27 the lifetime is typically $\tau \approx 10^{-15}$ s. This may be compared with the transit time $2R/v$ for the neutron, of speed v, to cross the nucleus, of diameter $2R$. The transit time is of order 10^{-21} s (problem 5.16 again) and so is many orders of magnitude less than the lifetime of the compound nucleus. Thus it is quite correct to regard the compound nucleus as a relatively long lived, independent, entity.

Finally, we may take an example from elementary particle physics. Figure 5.28 shows the cross section for scattering of a π^+ meson by a proton. The cross section is of the now familiar resonance shape, and its explanation is very similar to that of the resonances in figure 5.27. The π^+ meson and the proton form a relatively long-lived complex, analogous to the compound nucleus. The complex is simply called a *resonance*, and the width of the curve in figure 5.28 may be converted in the usual way into a lifetime of the resonance, which is in fact around 10^{-23} s. There are now many resonances known in high energy physics; the particular one shown in figure 5.28 is called the Δ^{++} resonance.

Worked example 5.5

Estimate the width in frequency of the curve for absorption of light by NaCl (figure 5.25) and hence estimate a lifetime for the optical mode. What, approximately, is the Q-value of the resonance?

Answer

Taking the width in wavelength as the width at 80 per cent transmission

$$\delta\lambda \approx 5\ \mu\text{m}$$

As in worked example 1.3, this is converted into a width in frequency

$$c = f\lambda$$

so

$$|\delta f| = \frac{f}{\lambda}\delta\lambda = \frac{c}{\lambda^2}\delta\lambda$$

The wavelength λ at the centre of the resonance is

$$\lambda = 61\ \mu\text{m}$$

so

$$|\delta f| = \frac{3 \times 10^8}{(6.1 \times 10^{-5})^2} \times 5 \times 10^{-5}$$

$$\approx 4 \times 10^{12}\ \text{Hz}$$

This gives for the lifetime

$$\delta t \approx \frac{1}{|\delta f|} = 2.5 \times 10^{-13} \text{ s}$$

The Q-value is

$$Q = \frac{\omega}{\delta\omega} = \frac{f}{|\delta f|} = \frac{\lambda}{\delta\lambda}$$

This gives

$$Q = \frac{61}{5} \approx 12$$

Problems

1. The restoring force of equation 5.1 may be represented by a spring which has tension Ku when its length is u. Show that the work done in extending the spring to length u_0 is $\frac{1}{2} Ku_0^2$, as stated below equation 5.7.

2. It was shown in the text.that the equation of motion (equation 5.14) follows from equation 5.11 for the conservation of energy. Show, by multiplying equation 5.14 by du/dt and integrating, that the converse is true.

3. A lightly damped oscillator has $\omega_1 = 10^7 \text{ s}^{-1}$ and $\Gamma = 10^6 \text{ s}^{-1}$. Find the subsequent motions with the following sets of initial conditions

$$\begin{array}{ll} u = 10^{-3} \text{ m} & du/dt = 0 \\ u = 0 & du/dt = 10^6 \text{ m s}^{-1} \\ u = 10^{-2} \text{ m} & du/dt = 10^6 \text{ m s}^{-1} \end{array}$$

4. A particle of mass m moves in the potential

$$V(x) = \frac{V_0(x^4 + 2a^4)}{a^2(x^2 + a^2)}$$

Sketch the potential, and find the points of stable and unstable equilibrium. Find the frequency of small oscillations about the points of stable equilibrium.

5. A particle of mass m moves in the potential

$$V(x) = V_0[\exp(\lambda x) + \exp(-\lambda x)]$$

Sketch the potential, identify the point of stable equilibrium, and find the frequency of small vibrations about the point of stable equilibrium.

6. The potential energy of an atom in a linear atomic chain has the form

$$V(x) = -\frac{a}{x} + \frac{b}{x^2}$$

where a and b are positive constants. Sketch the potential. Show that the equilibrium point is $x_0 = 2b/a$ and derive the angular frequency ω of small vibrations. If $a = 4 \times 10^{-30}$ J m, $b = 4 \times 10^{-40}$ J m^2, and the mass of the atom is 2×10^{-26} kg (these figures are roughly correct for something like sodium), calculate ω. Oscillations of large amplitude will be anharmonic. If the total energy of the oscillator is E (< 0), find the extreme points x_1 and x_2 of the oscillation. Define the mid-point of the oscillation as $\bar{x} = \frac{1}{2}(x_1 + x_2)$ and show that

$$\bar{x} = -a/2E$$

This result means that the mid-point moves out from $x_0 = 2b/a$ as $|E|$ decreases. What well-known thermal property of solids could be explained by this result?

7. Verify by direct substitution that $u = u_0 t \exp(-\frac{1}{2}\Gamma t)$ is a solution of the damped harmonic oscillator for critical damping, $b^2 = 4mK$.

8. The equation of motion of a damped harmonic oscillator is

$$m \frac{\mathrm{d}^2 x}{\mathrm{d}t^2} + b \frac{\mathrm{d}x}{\mathrm{d}t} + Kx = 0$$

with positive constants m, b and K. By substituting $x = A \exp(\lambda t)$, or otherwise, find the form of the solution. Sketch the solution for $b^2 < 4mK$ and for $b^2 > 4mK$.
Two coupled, undamped oscillators have the equations of motion

$$m_1 \frac{\mathrm{d}^2 x_1}{\mathrm{d}t^2} + k_{11} x_1 + k_{12} x_2 = 0$$

$$m_2 \frac{\mathrm{d}^2 x_2}{\mathrm{d}t^2} + k_{21} x_1 + k_{22} x_2 = 0$$

with all k's positive and $k_{11} k_{22} > k_{21} k_{12}$. A *normal mode* is defined as a motion in which x_1 and x_2 oscillate at the same frequency with a fixed relative phase. Derive expressions for the angular frequencies ω_a and ω_b of the two possible normal modes. What is the ratio x_1/x_2 in each normal mode? [UE]

9. The two solutions of the damped oscillator equation given in equation 5.42 can be written

$$u = u_0 \exp(-iz_1 t), \qquad u = u_0 \exp(-iz_2 t)$$

with

$$z_1 = \omega_1 - \tfrac{1}{2} i\Gamma$$
$$z_2 = -\omega_1 - \tfrac{1}{2} i\Gamma$$

Now (as a thought experiment) imagine the damping constant b to increase from zero. Plot the paths traced out by z_1 and z_2 in the complex z plane as b varies from 0 to ∞.

10. Show that the complex amplitude u_0 of the driven oscillator, equation 5.61, may be written

$$u_0 = R(\omega)F/m$$

with the *response function* $R(\omega)$ given by

$$R(\omega) = -\frac{1}{(\omega - z_1)(\omega - z_2)}$$

where z_1 and z_2 are the complex frequencies defined in the previous question. Note that if ω is regarded as a complex variable, $R(\omega)$ has simple poles at the vibration frequencies z_1 and z_2. This observation, that the poles of a response function occur at the natural oscillation frequencies of a system, is central to modern theoretical physics.

11. Confirm that the resonance function $P_r(\omega)$ of equation 5.81 has a maximum value of unity at $\omega = \omega_0$, as stated in the text. (Note that this result implies that the absorptive amplitude B (equation 5.64) does not have its maximum value exactly at $\omega = \omega_0$).

12. Show that the two positive-frequency solutions of equation 5.83, used in defining the Q-value, are

$$\omega_1 = (\omega_0{}^2 + \tfrac{1}{4}\Gamma^2)^{1/2} - \tfrac{1}{2}\Gamma$$
$$\omega_2 = (\omega_0{}^2 + \tfrac{1}{4}\Gamma^2)^{1/2} + \tfrac{1}{2}\Gamma$$

and hence that $\omega_2 - \omega_1 = \Gamma$, as quoted in equation 5.84.

13. In order to find the Q-value of the lorentzian line shape, (equation 5.87) the equation

$$P_L(\omega) = \tfrac{1}{2}$$

must be solved. Show that the two roots are

$$\omega_1 = \omega_0 - \tfrac{1}{2}\Gamma$$
$$\omega_2 = \omega_0 + \tfrac{1}{2}\Gamma$$

so that the Q-value is $Q = \omega_0/\Gamma$, the same as the Q-value for the exact resonance function $P_r(\omega)$.

14. What is the Q-value for the LCR circuit of figure 5.7?

15. Explain what is meant by the Q-value of a resonance. Give two examples of resonance in natural systems, and in each case suggest what dissipative processes may determine the Q-value. [UE]

16. As explained in the text, convert the width of a typical resonance in figure 5.27 into a lifetime τ of the corresponding compound nucleus. Compare the result with the transit time $2R/v$. (Take the nuclear radius R as 1 fm (= 10^{-15} m) The energy–momentum relation of the neutrons may be taken as the ordinary non-relativistic one.)

17. Find the lifetime of the Δ^{++} resonance of figure 5.28.

6

Equations for Waves

Many of the earlier sections of this book may be summarised by saying that they dealt with superposition properties of waves of various kinds. As we saw in section 1.2, if it is assumed that the equation of motion of the wave is linear, the amplitude of different waves may be added. This linear superposition has many interesting consequences, such as the interference pattern produced in the Young's slits experiment (figure 1.7), the distinction between phase and group velocity discussed in section 2.5, the various diffraction effects of chapter 3, and so on. However, there are some problems which cannot be tackled without explicit knowledge of the equation of motion itself. In particular, we always need the equation of motion when an expression for the *intensity* of the wave is required. The fact that the intensity is proportional to the square of the amplitude has been used at various points, but we did not know the constant of proportionality, which clearly depends on the properties of the medium through which the wave is travelling. Similarly, we need an equation of motion to deal with *interface problems* of various kinds. For example, in the microwave tunnelling experiment described in section 2.7, although it is clear on general grounds that there is a tunnel wave transmitted across the air gap, we cannot find its magnitude without knowing the equation of motion of the microwaves, and also the *boundary conditions* which apply at the interface between the dielectric and the air.

For the reasons given above, this last chapter is devoted to equations of motion for waves. Just two examples will be treated in detail; the equation for longitudinal acoustic waves in a solid, liquid or gas, and the equation for the 'particle waves' which we described in section 4.2. The equation for longitudinal acoustic waves is identical in form to that which can be derived for transverse acoustic waves, such as a wave on a string or a shear wave in a solid. The most notable omission from this chapter is the equation of motion for electromagnetic radiation. In order to derive this equation one requires a working knowledge of Maxwell's equations for the electromagnetic field, which in turn implies some background in elementary electricity and magnetism and an acquaintance with vector calculus. We must, therefore, regard the equation for electromagnetic waves as beyond our scope.

The displacement in a wave is a function of both position and time, $u = u(x, t)$. The equation of motion for u must obviously involve both x and t, and we shall see that it is written in terms of various partial derivatives of u with respect to x and t.

6.1 Wave Equation for Acoustic Waves

Consider a longitudinal wave travelling in the x-direction along an elastic bar. As shown rather schematically in figure 6.1, at any moment of time some portions

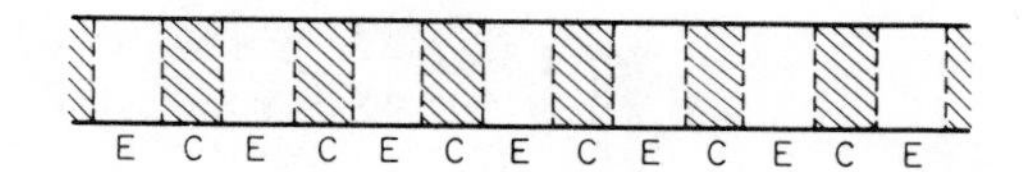

Figure 6.1 E extended regions. C compressed regions

of the bar are extended from their equilibrium length, and some are compressed. First of all we define $u(x, t)$ as the *longitudinal displacement* at position x and time t; that is, we take the cross-sectional area whose equilibrium position is at x, and we suppose that its position at time t is $x + u(x, t)$ (figure 6.2).

We now find the equation of motion of the element of the bar whose equilibrium position is between x and $x + \delta x$; as shown in figure 6.3, at time t this element lies between $x + u(x, t)$ and $x + \delta x + u(x + \delta x, t)$. As shown, the element is acted upon by the two forces F_1 and F_2. These forces are produced by the elastic properties of the bar, and their magnitudes depend on the amount by which the element has been extended or compressed. We make the usual assumption that the elastic force at any point is proportional to the amount of extension at that point. This is known as *Hooke's law*, and it will be recalled that it applies accurately for modest extensions. The extension of the element in question is the difference between the

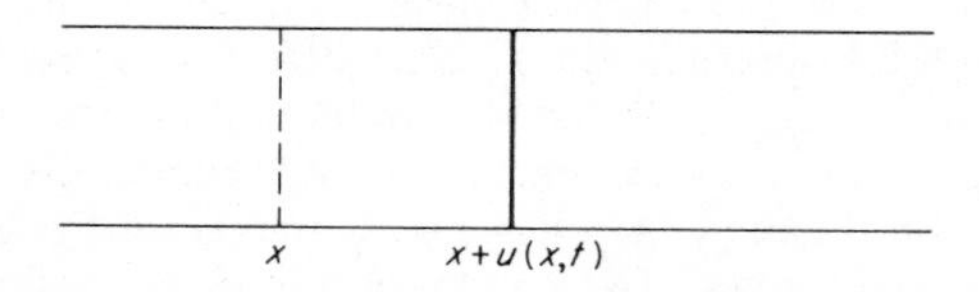

Figure 6.2 Displacement $u(x, t)$ in longitudinal wave

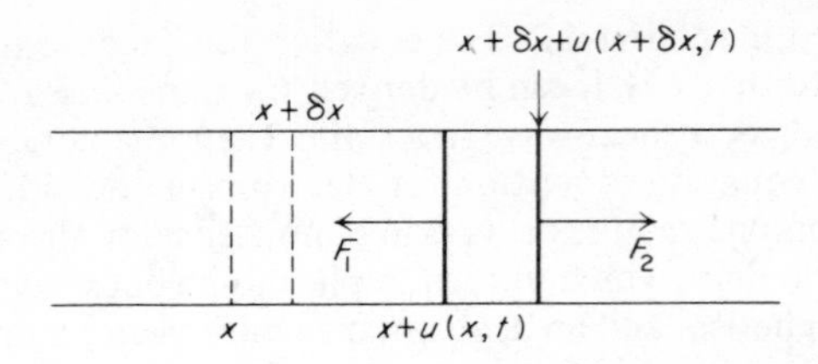

Figure 6.3 Forces on displaced element

strained and the unstrained length. The unstrained length is δx, and the strained length L_s is

$$L_s = x + \delta x + u(x + \delta x) - [x + u(x)] \tag{6.1}$$

Taking δx to be small we can expand $u(x + \delta x)$ to first order in δx

$$u(x + \delta x) = u(x) + \delta x \frac{\partial u}{\partial x} \tag{6.2}$$

whence

$$L_s = \delta x + \delta x \frac{\partial u}{\partial x} \tag{6.3}$$

The extension, therefore, is

$$L_s - \delta x = \delta x \frac{\partial u}{\partial x} \tag{6.4}$$

With the assumption that Hooke's law applies, the force F_1 is

$$F_1 = CA \frac{\text{extension}}{\text{unstrained length}} \tag{6.5}$$

and since the unstrained length is δx, this gives

$$F_1 = CA \frac{\partial u}{\partial x} \tag{6.6}$$

The cross-sectional area A has been introduced into the constant of proportionality; this has the advantage that the elastic constant C is the *Young's modulus* of the material of which the bar is made. If we were dealing with a wave in a bulk specimen, rather than in a bar, the appropriate elastic constant would be the *bulk modulus* rather than the Young's modulus. This difference arises because in a bar a longitudinal expansion is accompanied by a reduction in diameter, and there is no comparable effect in a bulk specimen. All our subsequent discussion could apply to bulk waves, on the understanding that C stands for the bulk modulus.

Equation 6.6 gives the force F_1. Any acceleration of the element of the bar is obviously due to the difference between the forces F_1 and F_2. Now F_1 involves the partial derivative $\partial u/\partial x$ evaluated for $u(x, t)$, while F_2 will have the same form with the derivative evaluated for $u(x + \delta x, t)$. Therefore

$$F_2 - F_1 = CA\left[\frac{\partial u(x + \delta x, t)}{\partial x} - \frac{\partial u(x, t)}{\partial x}\right] \tag{6.7}$$

We write the term in square brackets as

$$\delta x \frac{\partial}{\partial x}\frac{\partial u}{\partial x} = \delta x \frac{\partial^2 u}{\partial x^2}$$

by the definition of the second derivative $\partial^2 u/\partial x^2$. The final form for the net force on the element is therefore

$$F_2 - F_1 = CA\delta x \frac{\partial^2 u}{\partial x^2} \tag{6.8}$$

The equation of motion may now be found by equating the net force to the mass of the element times its acceleration a. The mass is $\rho A \delta x$ where ρ is the density of the unstrained bar, since δx is the unstrained length. The acceleration is

$$\begin{aligned} a &= \frac{\mathrm{d}^2}{\mathrm{d}t^2}[x + u(x, t)] \\ &= \frac{\partial^2 u}{\partial t^2} \end{aligned} \tag{6.9}$$

where the second line follows since x is simply the fixed initial position of the element. Collecting together these expressions gives for the equation of motion

$$CA\delta x \frac{\partial^2 u}{\partial x^2} = \rho A \delta x \frac{\partial^2 u}{\partial t^2} \tag{6.10}$$

or simply

$$\frac{1}{v^2}\frac{\partial^2 u}{\partial t^2} = \frac{\partial^2 u}{\partial x^2} \tag{6.11}$$

with

$$v^2 = C/\rho \tag{6.12}$$

The step from equation 6.10 to equation 6.11 involves cancelling the cross-sectional area A and the length δx, both of which might be expected to disappear from the equation of motion of the element.

Equation 6.11 is the result required, since it relates the time and space dependence of the displacement u. The next task is to find the general form of solutions of the equation. We shall see that it describes non-dispersive propagation, with the velocity v at any frequency given by equation 6.12. To begin with, it is worth while finding the velocity of propagation of a plane wave of frequency ω. Substituting into equation 6.11 the trial solution

$$u(x, t) = u_0 \exp i(kx - \omega t) \tag{6.13}$$

gives for the partial derivatives

$$\frac{\partial u}{\partial x} = iku(x, t) \tag{6.14}$$

$$\frac{\partial^2 u}{\partial x^2} = -k^2 u(x, t) \tag{6.15}$$

and likewise

$$\frac{\partial^2 u}{\partial t^2} = -\omega^2 u(x, t) \tag{6.16}$$

Substituting these into equation 6.11 we find

$$\frac{\omega^2}{v^2} u(x, t) = k^2 u(x, t) \tag{6.17}$$

Thus equation 6.13 is a solution, provided that

$$\omega = vk \tag{6.18}$$

and we see that v is the velocity of propagation of the plane wave. Notice that v is independent of the frequency ω, so that we are indeed dealing with non-dispersive propagation. This result is consistent with the dispersion equation, equation 2.33, for wave propagation on the 'mass and spring' lattice of figure 2.23. As we pointed out in discussing equation 2.33, in the long-wavelength limit the 'grainy' structure of the medium is irrelevant, and the medium may be treated as an elastic continuum. Equation 6.11 is derived from a continuum approximation, and so it naturally gives the non-dispersive propagation which is the small-k limit of equation 2.33.

Equation 6.11 is usually called the wave equation, although as we have just seen it only describes non-dispersive waves. The single-frequency wave of equation 6.13 is a solution of the wave equation provided $\omega = vk$, but it is clearly not the general solution. It will now be shown that a more general solution is

$$u(x, t) = f(x - vt) + g(x + vt) \tag{6.19}$$

where f and g are arbitrary (twice differentiable) functions. In fact equation 6.19 is the most general solution, but this will not be proved.

We may easily verify that $f(x - vt)$ is a solution of the wave equation. We write

$$u(x, t) = f(y) \tag{6.20}$$

with

$$y = x - vt \tag{6.21}$$

Then

$$\frac{\partial u}{\partial x} = f'(y) \tag{6.22}$$

where $f'(y)$ is the first derivative with respect to y, and similarly

$$\frac{\partial u}{\partial t} = \frac{\mathrm{d}f}{\mathrm{d}y}\frac{\partial y}{\partial t} = -vf'(y) \tag{6.23}$$

The second derivatives are

$$\frac{\partial^2 u}{\partial x^2} = f''(y) \tag{6.24}$$

and

$$\begin{aligned}\frac{\partial^2 u}{\partial t^2} &= -v\frac{\partial}{\partial t}f'(y) = -v\frac{\mathrm{d}f'}{\mathrm{d}y}\frac{\partial y}{\partial t}\\ &= v^2 f''(y)\end{aligned} \tag{6.25}$$

Combining equations 6.24 and 6.25 gives

$$\frac{1}{v^2}\frac{\partial^2 u}{\partial t^2} = \frac{\partial^2 u}{\partial x^2} \tag{6.26}$$

which is the result required. The proof that $g(x + vt)$ is a solution is similar, and since the wave equation is linear in u, we see that equation 6.19 is also a solution.

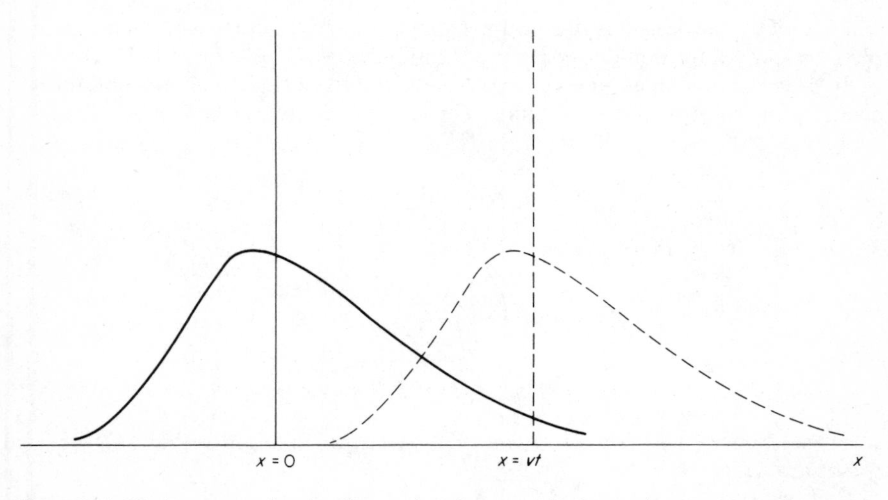

Figure 6.4 Displacement $u(x, t) = f(x - vt)$ at time $t = 0$ (full curve) and at time t (dashed curve)

It is easy to see what the solution $u(x, t) = f(x - vt)$ represents. If the displacement is drawn as a function of x at time $t = 0$ we have simply

$$u(x, 0) = f(x) \tag{6.27}$$

that is, the function $f(x)$. At time t, the displacement has the same shape, but is centred on the point $x = vt$ rather than on the origin $x = 0$ (figure 6.4). We therefore have a displacement with the shape $f(x)$ travelling to the right with velocity v. In a similar way, it can be shown that $g(x + vt)$ represents a displacement of shape $g(x)$ travelling to the left with velocity v.

The result just obtained is not surprising. We saw previously, in discussing figure 2.25, that in a dispersive medium a general disturbance changes shape as it propagates. It is a consequence of our discussion at that point, although not brought out explicitly there, that a general disturbance does not change shape as it propagates through a non-dispersive medium. The fact that equation 6.19 is a solution of the wave equation is therefore equivalent to the fact that the wave equation has the simple dispersion relation of equation 6.18.

6.2 Intensity and Wave Impedance

As pointed out at the beginning of this chapter one reason for deriving the wave equation is that we can then find an expression for the intensity of the wave. The intensity I is defined as the rate of energy transfer across unit area. Writing $\langle E \rangle$ for the energy stored per unit volume of the medium (note that $\langle E \rangle$ has the dimensions energy L^{-3}), we have

$$I = v\langle E \rangle \tag{6.28}$$

where strictly speaking v is the group velocity, which of course is the same as the phase velocity v for non-dispersive propagation.

In order to find an expression for $\langle E \rangle$, we return to figure 6.3 and find the energy of an element of the medium. Then we restrict our attention to a single-frequency wave

$$u(x, t) = u_0 \cos(kx - \omega t) \tag{6.29}$$

The kinetic energy T of the element δx is

$$\begin{aligned} T &= \tfrac{1}{2}\rho A \delta x \left(\frac{\partial u}{\partial t}\right)^2 \\ &= \tfrac{1}{2}\rho A \delta x u_0^2 \omega^2 \sin^2(kx - \omega t) \end{aligned} \tag{6.30}$$

since the element has mass $\rho A \delta x$ and velocity $\partial u/\partial t$. In order to find $\langle E \rangle$ we need the time average $\langle T \rangle$

$$\langle T \rangle = \tfrac{1}{4}\rho A \delta x u_0^2 \omega^2 \tag{6.31}$$

by the usual rule. The potential energy V of the element is given by

$$V = \tfrac{1}{2} CA\delta x \left(\frac{\text{extension}}{\text{unstrained length}}\right)^2$$

$$= \tfrac{1}{2} CA\delta x \left(\frac{\partial u}{\partial x}\right)^2$$

$$= \tfrac{1}{2} CA\delta x u_0^2 k^2 \sin^2 (kx - \omega t) \tag{6.32}$$

The time average is therefore

$$\langle V \rangle = \tfrac{1}{4} CA\delta x u_0^2 k^2 \tag{6.33}$$

Note that since $Ck^2 = \rho\omega^2$, from equations 6.12 and 6.18 for the velocity v, the time averages $\langle T \rangle$ and $\langle V \rangle$ are equal. The energy stored per unit volume is equal to $\langle T \rangle + \langle V \rangle$ divided by the volume $A\delta x$ of the element

$$\langle E \rangle = \frac{\langle T \rangle + \langle V \rangle}{A\delta x} = \frac{2\langle T \rangle}{A\delta x} = \tfrac{1}{2} \rho u_0^2 \omega^2 \tag{6.34}$$

Equation 6.28 then gives for the intensity

$$I = \tfrac{1}{2} \rho u_0^2 \omega^2 v \tag{6.35}$$

which is the principal result of this section.

With an eye to later developments, we may rewrite equation 6.35 in terms of the *specific acoustic impedance* Z, defined as

$$Z = \rho v = (\rho C)^{1/2} \tag{6.36}$$

The intensity may then be written

$$I = \tfrac{1}{2} Z u_0^2 \omega^2 = \tfrac{1}{2} Z s_0^2 \tag{6.37}$$

where s_0 is the maximum velocity $u_0\omega$ of the element. We shall see that the impedance is important in dealing with transmission and reflection at an interface, and in addition it makes the analogies between acoustic and electromagnetic waves particularly clear.

6.3 Radiation Pressure

We have just seen that a travelling wave transmits energy. Consequently if a wave falls on an absorbing or reflecting surface it exerts a pressure on the surface. The following virtual work argument yields a simple expression for this *radiation pressure.*

First suppose the wave falls on an absorbing surface of area A, and let the radiation pressure (force per unit area) on the surface be P. The absorber is held in place by a reaction force PA equal and opposite to the force due to the radiation pressure. If the absorber is moved a distance δx towards the source of

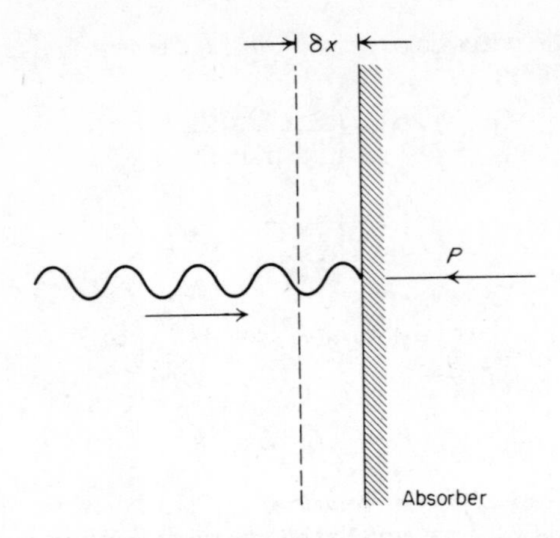

Figure 6.5 Radiation pressure on an absorber

the wave (figure 6.5) then the reaction force does work $PA\delta x$. At the same time the wave no longer occupies a volume $A\delta x$, so the energy of the wave is reduced by $\langle E \rangle A\delta x$. Equating this to the work done by the reaction force gives

$$PA\delta x = \langle E \rangle A\delta x \tag{6.38}$$

or simply

$$P = \langle E \rangle \tag{6.39}$$

which is the expression for the radiation pressure on an absorbing surface.

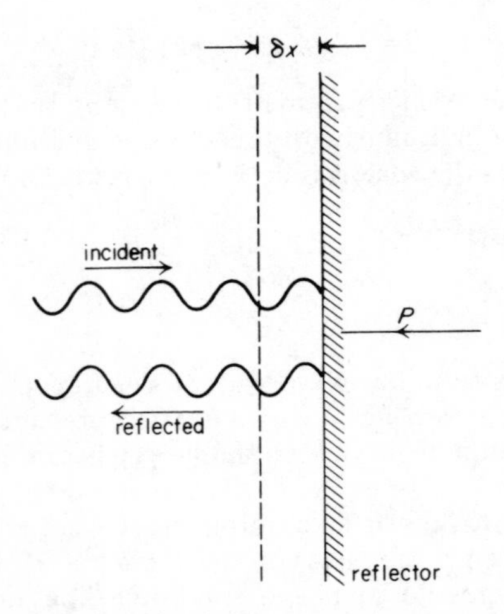

Figure 6.6 Radiation pressure on a reflector

If instead of an absorber, we have a reflecting surface, there is a reflected wave as well as an incident wave (figure 6.6). The energy removed from the wave is therefore $2\langle E \rangle A\delta x$ and so

$$P = 2\langle E \rangle \tag{6.40}$$

for a reflector. It should be noted that equations 6.39 and 6.40 are (as they must be) dimensionally correct. The left-hand side is force per unit area, and the right-hand side is energy per unit volume.

6.4 Reflection and Transmission at Interfaces

There is obviously a general class of problems concerning the transmission of waves out of one medium into another. A musician, for example, sets up a note as a standing wave on a string, in a pipe, or whatever, and the wave then propagates from his instrument into the surrounding air. Again, the transmission and reception of a radio signal involves several stages of transfer between different media. We shall study in detail the simplest example of a problem of this kind, namely the transmission and reflection of a wave at an abrupt plane interface between two acoustic media. Simple as the example is, it will enable us to draw some important general conclusions.

Consider, as shown in figure 6.7, an interface between two media, of Young's modulus and density C_1, ρ_1 and C_2, ρ_2 respectively. The displacement $u(x, t)$ in either medium satisfies the wave equation. In order to relate the waves on either side, we must derive boundary conditions relating the displacements u_1 and u_2 at $x = 0$. The first condition is

$$u_1(0, t) = u_2(0, t) \tag{6.41}$$

since otherwise the two media would no longer be contiguous. A second boundary condition can be derived by considering the forces acting on a small element, from $x = -\delta$ to $x = +\delta$, at the interface. The argument leading to equation 6.6 shows that the net force per unit area acting on this element is

$$F = C_2 \frac{\partial u_2}{\partial x} - C_1 \frac{\partial u_1}{\partial x} \tag{6.42}$$

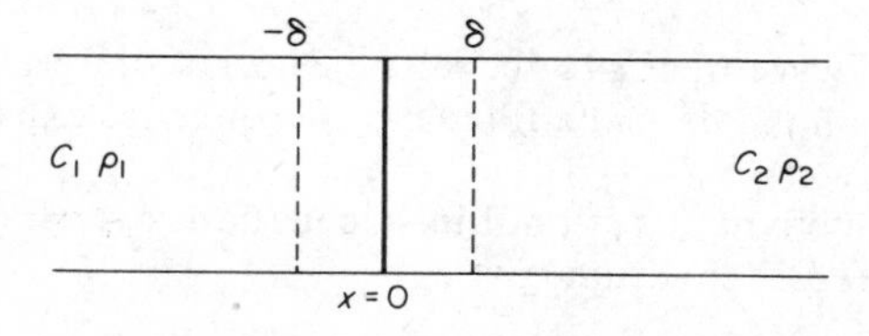

Figure 6.7 An abrupt interface between two media

However, the mass of this element can be made arbitrarily small by making δ small, and since the element cannot have infinite acceleration, the net force must vanish. Thus the second boundary condition is

$$C_1 \frac{\partial u_1}{\partial x} = C_2 \frac{\partial u_2}{\partial x} \tag{6.43}$$

We now restrict our attention to a wave of a single frequency ω. Since the wave velocities are different in the two media, we have different wave numbers k_1 and

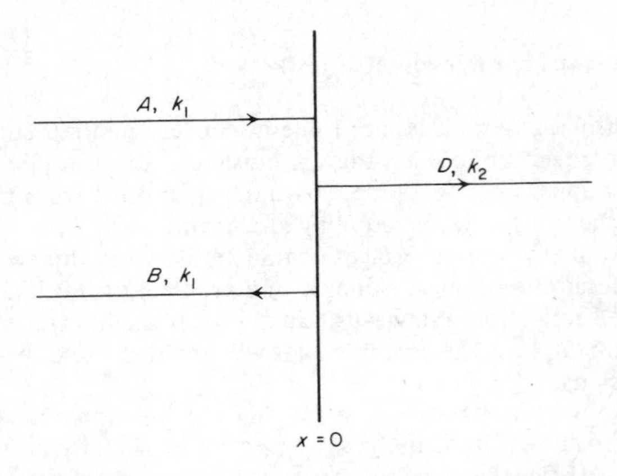

Figure 6.8 Incident, reflected and transmitted waves

k_2. Suppose we have an incident wave in medium 1 of amplitude A. The quantities of interest are the amplitude D of the transmitted wave in medium 2, and the reflected amplitude B in medium 1. It is apparent from figure 6.8 that the total displacements in the two media are

$$u_1(x, t) = A \exp i(k_1 x - \omega t) + B \exp [-i(k_1 x + \omega t)] \tag{6.44}$$

$$u_2(x, t) = D \exp i(k_2 x - \omega t) \tag{6.45}$$

with the wave numbers given by

$$k_1 = \omega/v_1 \tag{6.46}$$

$$k_2 = \omega/v_2 \tag{6.47}$$

Since the reflected wave travels to the left, the corresponding exponential involves $k_1 x + \omega t$ while retaining the same frequency dependence $\exp(-i\omega t)$ as the incident wave.

We now apply the boundary conditions, equations 6.41 and 6.43, to equations 6.44 and 6.45. The first condition gives

$$A + B = D \tag{6.48}$$

since we are matching the waves at $x = 0$, and we have dropped the factor $\exp(-i\omega t)$ throughout. The second boundary condition gives

$$C_1 k_1 A - C_1 k_1 B = C_2 k_2 D \qquad (6.49)$$

It is convenient to rewrite the second condition in terms of the impedance Z which we introduced in equation 6.36.

$$C_1 k_1 = C_1 \omega / v_1 = \omega \sqrt{(C_1 \rho_1)} = \omega Z_1 \qquad (6.50)$$

where equation 6.12 is used for the velocity. Equation 6.49 therefore becomes

$$Z_1 (A - B) = Z_2 D \qquad (6.51)$$

It is now easy to find D and B from equations 6.48 and 6.51

$$D = TA \qquad (6.52)$$

$$B = RA \qquad (6.53)$$

where the transmission coefficient T and reflection coefficient R are given by

$$T = \frac{2}{1 + Z_2/Z_1} \qquad (6.54)$$

$$R = \frac{1 - Z_2/Z_1}{1 + Z_2/Z_1} = \frac{Z_1 - Z_2}{Z_1 + Z_2} \qquad (6.55)$$

It is worth noting that T and R depend only on the impedances Z_1 and Z_2, and in fact only on the *relative impedance* Z_2/Z_1. We can now see that the impedance is a very important parameter occurring as it does both in the intensity, equation 6.37, and in the transmission and reflection coefficients just derived. The two occurrences are related. There is no absorption of energy at the barrier in figure 6.8, so the power arriving at the barrier must equal the power leaving it. In other words, the intensity in the incident beam must equal the sum of the intensities in the reflected and transmitted waves. This requires

$$Z_1 A^2 = Z_1 B^2 + Z_2 D^2 \qquad (6.56)$$

or in terms of T and R,

$$Z_1 = Z_1 R^2 + Z_2 T^2 \qquad (6.57)$$

It is a straightforward matter to verify that the expressions for T and R do satisfy this identity.

Equations 6.54 and 6.55 call for further discussion. In particular, it can be seen that in order to maximise the transmission at an interface, or equivalently to minimise the reflection, one must arrange to have Z_1 and Z_2 as nearly equal as possible. Arranging for a minimum of reflection between two media is known as *impedance matching* of the media, and is obviously an important practical problem. Before dealing with impedance matching, however, we shall digress to state the properties of electromagnetic radiation which parallel those we have derived for acoustic waves.

Worked example 6.1

Sound is incident from a medium 1, say air, of fixed impedance Z_1 on an interface with a medium of impedance Z_2. Sketch the variation with Z_2/Z_1 of the transmissior and reflection coefficients T and R.

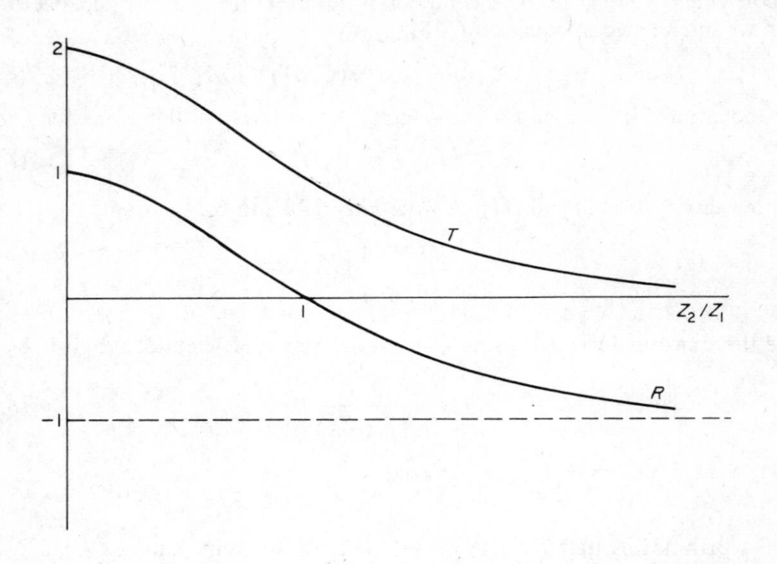

Answer

The coefficients are given by equations 6.54 and 6.55

$$T = \frac{2}{1+x}$$

$$R = \frac{1-x}{1+x}$$

where $x = Z_2/Z_1$. For $x = 0$, $T = 2$. As x increases T decreases, and approaches zero (like $1/x$) as $x \to \infty$. For $x = 0$, $R = 1$. At $x = 1$, $R = 0$ (and in fact $T = 1$)—this is impedance matching. As $x \to \infty$, $R \to -1$. The coefficients are therefore as sketched above.

The transmission coefficient $T = 2$ for $Z_2 = 0$ may seem surprising at first sight. However, the transmitted intensity is proportional to $Z_2 T^2$ and is zero for $Z_2 = 0$.

6.5 Electromagnetic Waves

As mentioned earlier we shall not derive the equation for electromagnetic waves. However, the main results are very similar to those found for acoustic waves, and are therefore worth stating. As we know, the velocity of the waves is given by

$$v = c/n \tag{6.58}$$

where $c = 3 \times 10^8$ m s^{-1} is the velocity of light in a vacuum, and n is the refractive index. The displacement in the wave may be specified as the electric field amplitude E_0, measured in volts metre^{-1}, or as the magnetic field amplitude H_0 measured in amps metre^{-1}. The two are related by the *impedance* of the medium

$$E_0 = H_0 Z \tag{6.59}$$

where Z is measured in ohms. In a vacuum, the value of the impedance is

$$Z_0 = 377\Omega \tag{6.60}$$

and in a general, non-magnetic, dielectric

$$Z = Z_0/n \tag{6.61}$$

where n is the refractive index.

As in an acoustic wave, the impedance is involved in the expressions for both intensity and reflection and transmission coefficients. The intensity is

$$I = \tfrac{1}{2}E_0^2/Z = \tfrac{1}{2}ZH_0^2 \tag{6.62}$$

The transmission and reflection coefficients may be written in terms of either the electric field or the magnetic field. In terms of the electric field, the expressions are

$$T_E = \frac{2}{1 + Z_2/Z_1} \tag{6.63}$$

$$R_E = \frac{Z_2 - Z_1}{Z_2 + Z_1} \tag{6.64}$$

Except for a difference of sign in the reflection coefficient, these are the same as the expressions for an acoustic wave, equations 6.54 and 6.55. Using equation 6.61, we can express these coefficients in terms of the refractive indices of the two media

$$T_E = \frac{2}{1 + n_1/n_2} \tag{6.65}$$

$$R_E = \frac{n_1 - n_2}{n_1 + n_2} \tag{6.66}$$

Worked example 6.2

Sunshine has an intensity of 1.35×10^3 W m^{-2}. Find the amplitudes E_0 and H_0. What is the radiation pressure on a mirror reflecting sunshine?

Answer

From equation 6.62, $E_0 = (2IZ_0)^{1/2}$ and $H_0 = (2I/Z_0)^{1/2}$, where $Z_0 = 377\ \Omega$ is the impedance of free space. Putting in the numerical values, we find

$$E_0 = 1000 \text{ V m}^{-1}$$

$$H_0 = 2.68 \text{ A m}^{-1}$$

The radiation pressure on a reflecting surface is $P = 2\langle E \rangle$ (equation 6.40), or using equation 6.28 for the intensity, $P = 2I/c$. This gives

$$P = 9 \times 10^{-6} \text{ N m}^{-2}$$

6.6 Impedance Matching

Many applications involve transfer of a wave from one medium to another. For example, the transmission and reception of a radio signal involves transfer from the transmitting aerial to free space, and again from free space to the receiving aerial. Another example is afforded by any musical instrument; the sound produced say in an organ pipe has to be transmitted into the space surrounding the pipe. It can be seen from equation 6.55 for the reflection coefficient that unless the impedances Z_1 and Z_2 are fairly close in value, most of the energy incident upon an abrupt interface will be reflected rather than transmitted. This section describes three general techniques of impedance matching by means of which reflections can be minimised.

In some cases, it is possible to adjust the parameters of one medium so that its impedance is close to that of the other. Thus a radio aerial is designed so that its impedance matches that of free space.

If the media themselves cannot be manipulated, it is possible to cut down reflections by designing a gradual interface rather than an abrupt one. The flared out horn at the end of a musical instrument like a trumpet has the effect of

Figure 6.9 Impedance matching by flared horn (*Crown copyright*)

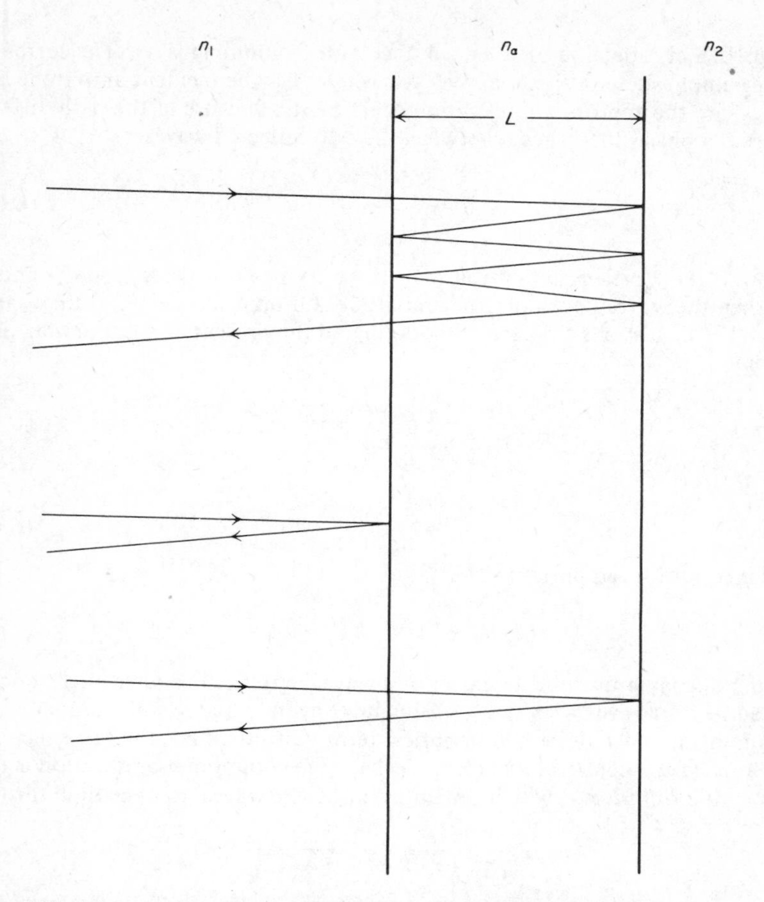

Figure 6.10 A bloomed interface with possible paths for reflected light. For clarity, the paths for a small, non-zero angle of incidence are shown, but the calculation is restricted to normal incidence.

improving the transfer of power between the inside of the instrument and the surrounding air (figure 6.9).

A third possibility is to *bloom* a sharp interface, as is done with camera lenses. Suppose as in figure 6.10 we interpose a layer of thickness L with refractive index n_α between medium 1 in which the light is incident and medium 2, into which we wish to transmit as much light as possible. For example, medium 1 might be air, and medium 2 a lens. We maximise the transmission into 2 by minimising reflection back into 1. The calculation is simplified by ignoring light paths involving multiple reflection like that shown at the top of figure 6.10. This is obviously a valid approximation if the reflection coefficient is small, that is, if the impedances of the three media are all fairly similar in magnitude. There is no difficulty, in principle, in making a calculation which allows for multiple reflections. The two possible paths for reflected light are then the direct reflection, in the middle of figure 6.10, and reflection off the back face of the blooming layer. The corresponding amplitudes are $R_{1\alpha}A$ and $R_{\alpha 2}A$, where A is the incident amplitude and $R_{1\alpha}, R_{\alpha 2}$, are the appropriate reflection coefficients. Because of the path difference $2L$ there is a phase difference δ between the two reflected waves

$$\delta = \frac{2L}{\lambda_\alpha} 2\pi \tag{6.67}$$

where λ_α is the wavelength in medium α. If we arrange to have $R_{1\alpha} = R_{\alpha 2}$ and $\delta = \pi$, then the waves have the same amplitude and opposite phase, so they cancel exactly. The conditions for perfect blooming, in the present approximation, are therefore

$$\frac{n_1 - n_\alpha}{n_1 + n_\alpha} = \frac{n_\alpha - n_2}{n_\alpha + n_2} \tag{6.68}$$

and

$$L = \lambda_\alpha/4 \tag{6.69}$$

The first condition simplifies to

$$n_\alpha = (n_1 n_2)^{1/2} \tag{6.70}$$

The blooming layer must be a quarter of a wavelength thick, and as might be expected its refractive index takes a value intermediate between n_1 and n_2.

Blooming has been described in optical terms, but of course the same technique can be applied to acoustical matching. Perhaps the commonest application is the matching of a *transducer*, which generates ultrasonic waves, to a medium through

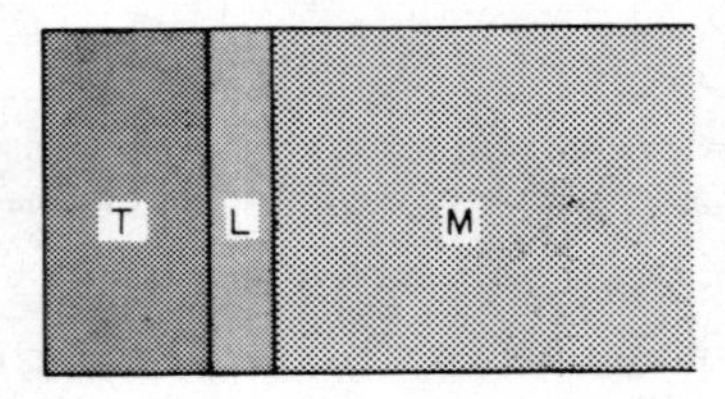

Figure 6.11 Transducer T, matching layer L, and medium M

which the waves must travel. As sketched in figure 6.11, a matching layer of thickness $\lambda/4$ and impedance $Z_L = (Z_T Z_M)^{1/2}$ ensures perfect transfer in the approximation that multiple reflections are ignored.

Worked example 6.3

Calculate the thickness and refractive index of a blooming layer for light of wavelength 500 nm normally incident from air on to a glass lens of refractive index 1.5.

Answer
The refractive index is given by equation 6.70

$$n_\alpha = (1.5)^{1/2} = 1.22$$

The wavelength in the blooming layer is equal to the free-space wavelength divided by the refractive index

$$\lambda_\alpha = 500/1.22 = 408 \text{ nm}$$

The thickness required is $\lambda_\alpha/4 \approx 100$ nm.

6.7 The Schrödinger Equation

We saw at the end of section 4.3 that the basic problem of quantum mechanics is to find an equation for the wave function of a particle moving in a variable potential. The appropriate equation is called the Schrödinger equation, after its inventor. It must be emphasised that Schrödinger's equation is a step forward, and cannot in any way be derived; the evidence for its validity is the very wide agreement of its predictions with experimental results. A perfectly rigorous way of proceeding would be to postulate the equation and derive some consequences. However, that approach is rather austere, and so we start with some remarks aimed at making Schrödinger's equation seem plausible.

We restrict our attention to differential equations for the wave function $\psi(x, y, z, t)$, so that we have to deal with partial derivatives of ψ with respect to time and the spatial co-ordinates x, y and z. We shall generally consider motion in one dimension, with one spatial co-ordinate x. To begin with, if the potential function $V(x)$ is zero, we expect to find plane wave solutions with a definite wave number and frequency

$$\psi(x, t) = \psi_0 \exp i(kx - \omega t) \tag{6.71}$$

We know from the experimental evidence presented in chapter 4 that this plane wave corresponds to energy $E = \hbar\omega$ and momentum $p = \hbar k$. Furthermore, for a particle we have, as in equation 4.20

$$E = \frac{p^2}{2m} \tag{6.72}$$

This is equivalent to

$$\hbar\omega = \frac{\hbar^2 k^2}{2m} \tag{6.73}$$

We therefore expect Schrödinger's equation for $V(x) = 0$ to be such that equation

6.71 is a solution on condition that equation 6.73 holds. Now the partial derivatives of the plane wave $\psi(x, t)$ of equation 6.71 are

$$\frac{\partial \psi}{\partial t} = -i\omega\psi \tag{6.74}$$

$$\frac{\partial \psi}{\partial x} = ik\psi \tag{6.75}$$

$$\frac{\partial^2 \psi}{\partial x^2} = -k^2\psi \tag{6.76}$$

The simplest differential equation which will produce equation 6.73 as a condition is therefore

$$i\hbar \frac{\partial \psi}{\partial t} = -\frac{\hbar^2}{2m}\frac{\partial^2 \psi}{\partial x^2} \tag{6.77}$$

since on substitution of equations 6.74 and 6.76 it is found that the plane wave is a solution if

$$\hbar\omega\psi_0 \exp i(kx - \omega t) = \frac{\hbar^2 k^2}{2m} \psi_0 \exp i(kx - \omega t) \tag{6.78}$$

After removal of common factors, equation 6.73 is recovered.

Equation 6.77 is Schrödinger's equation for the case when $V(x) = 0$. It may be obtained from the equation for conservation of energy, equation 6.72, by the following rule:

Multiply the energy conservation equation by ψ. Then convert E and p to differential operators by the prescriptions

$$E \to i\hbar \frac{\partial}{\partial t} \tag{6.79}$$

$$p \to -i\hbar \frac{\partial}{\partial x} \tag{6.80}$$

The resulting equation is Schrödinger's equation.

What should be done when $V(x)$ is non-zero? First, suppose that $V(x)$ is varying very slowly, so that it is more or less constant over many wavelengths of ψ (figure 6.12). The equation for conservation of energy is

$$E = \frac{p^2}{2m} + V(x) \tag{6.81}$$

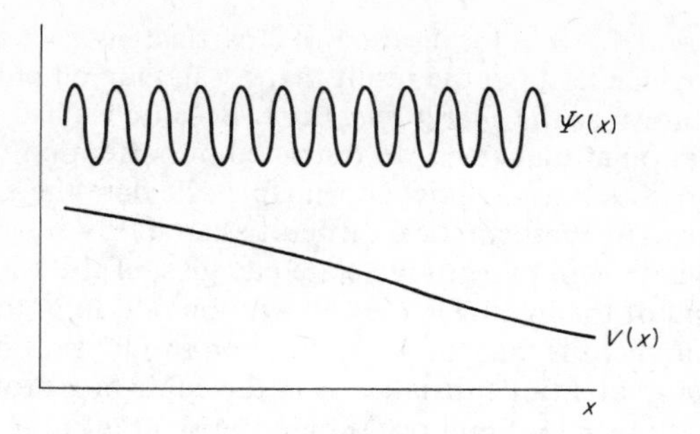

Figure 6.12 $V(x)$ slowly varying

$V(x)$ is more or less constant for a wide range round any particular point x, and may be regarded simply as a shift of the zero of E

$$E - V(x) = \frac{p^2}{2m} \tag{6.82}$$

Application of the basic rule then gives the differential equation

$$i\hbar \frac{\partial \psi}{\partial t} = -\frac{\hbar^2}{2m}\frac{\partial^2 \psi}{\partial x^2} + V(x)\psi \tag{6.83}$$

This equation holds when $V(x)$ is slowly varying. The general case, when $V(x)$ is not slowly varying, is dealt with simply by postulating that equation 6.83 continues to hold. As we pointed out, the evidence in support of this postulate is that it gives the correct predictions for such things as the hydrogen energy level scheme of equation 4.22 and figure 4.7. Equation 6.83 then, is the general form of Schrödinger's equation, and it is obtained from the energy conservation equation by the general rules we have stated for conversion of $E\psi$ and $p\psi$ to partial derivatives.

The formal framework of quantum mechanics has now been almost completely derived. The main addition that is required is that in order to describe all the interference and diffraction effects correctly, ψ must be taken as a complex function. In consequence, the interpretation we gave at the end of section 4.4 is modified slightly. The probability that a particle will be detected at a point x is equal to the square of the modulus $|\psi(x, t)|^2 = \psi^*(x, t)\psi(x, t)$.

In most problems, we deal with a particle of fixed total energy E. This gives a single frequency $\omega = E/\hbar$, and the time dependence of the wave function is simply $\exp(-i\omega t)$. Then we may write

$$\psi(x, t) = \exp(-i\omega t)\phi(x) \tag{6.84}$$

and substituting this into equation 6.83, we find that $\phi(x)$ satisfies

$$-\frac{\hbar^2}{2m}\frac{d^2\phi}{dx^2} + V(x)\phi = E\phi \tag{6.85}$$

which is known as the *time-independent Schrödinger equation.*

We saw in section 4.3 that localisation of a particle involves a standing-wave condition on ψ, which leads to the result that E can take on only one of a discrete range of values known as the *energy spectrum.* Because we had not found the Schrödinger equation at that stage, we restricted our attention to the cubic box potential of figure 4.9. We may now in principle calculate the energy spectrum for varying potentials. The mathematical difficulties are fairly substantial, and so we shall simply calculate here two ground-state energies; of the harmonic oscillator in section 6.8, and of the hydrogen atom in section 6.9. In both these cases the localisation requirement is that the wave function should vanish at infinity, not that it should vanish at a finite distance as in the cubic box problem. However, the condition that the wave functions should vanish at infinity is still a standing wave condition of a kind, and as we shall see it gives discrete energy values.

6.8 The Quantum Mechanical Harmonic Oscillator

Consider the motion of a particle of mass m in the very familiar oscillator potential

$$V(x) = \tfrac{1}{2}Kx^2 \tag{6.86}$$

shown in figure 6.13. As we saw in chapter 5, the simple harmonic oscillator is very important, because it describes all small vibrations about equilibrium. The quantum mechanics of the harmonic oscillator is therefore of fundamental importance, since it gives the quantum mechanical description of all small vibrations.

With the potential of equation 6.86, the time-independent Schrödinger equation (equation 6.85) becomes

$$-\frac{\hbar^2}{2m}\frac{\mathrm{d}^2\phi}{\mathrm{d}x^2} + \tfrac{1}{2}Kx^2\phi = E\phi \tag{6.87}$$

This is an equation to determine the energy E. It may be expected that as in the 'cubic box' problem there is a whole set of solutions, rather as sketched in figure 4.11, and that to each possible standing wave there is an allowed energy E. We

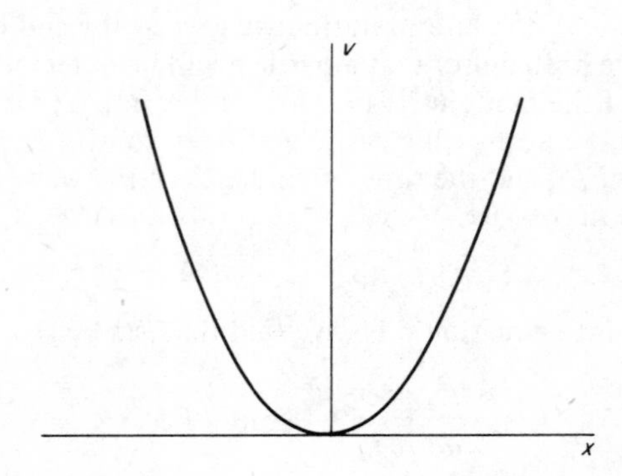

Figure 6.13 Harmonic oscillator potential

shall simply find the lowest, or ground state, energy E_0. An approximate value for E_0 was found in section 4.4 by using the uncertainty principle. It was argued there that if the wave function is highly localised then the kinetic energy T is large, while if the wave function is very widespread then the potential energy V is large. The correct degree of localisation is that which minimises $T + V$.

Equation 6.87 may be solved by writing $\phi(x)$ as a power series in x, and determining the coefficients so that the differential equation is satisfied. Details of this method are given in quantum mechanics textbooks. For the ground-state energy, a simpler approach may be used. It may be expected that the ground-state wave function is symmetrical about $x = 0$, as sketched in figure 4.16. This holds for the cubic box ground-state function, equation 4.24 and figure 4.11, and in fact it holds for the ground-state function in any symmetrical potential. We therefore assume that $\phi(x)$ is a function of x^2 only and put

$$\phi(x) = f(y) \tag{6.88}$$

with

$$y = x^2 \tag{6.89}$$

The usual rule for differentiating a function of a function gives

$$\frac{d\phi}{dx} = \frac{df}{dy}\frac{dy}{dx} = 2x\frac{df}{dy} \tag{6.90}$$

and

$$\begin{aligned}\frac{d^2\phi}{dx^2} &= 2\frac{df}{dy} + 2x\frac{d}{dx}\frac{df}{dy}\\ &= 2\frac{df}{dy} + 2x\frac{d^2f}{dy^2}\frac{dy}{dx}\\ &= 2\frac{df}{dy} + 4x^2\frac{d^2f}{dy^2}\\ &= 2\frac{df}{dy} + 4y\frac{d^2f}{dy^2}\end{aligned} \tag{6.91}$$

Substitution into equation 6.87 yields

$$-\frac{\hbar^2}{2m}\left(2\frac{df}{dy} + 4y\frac{d^2f}{dy^2}\right) + \tfrac{1}{2}Kyf = Ef \tag{6.92}$$

After some scrutiny it can be seen that the simplest function which satisfies this equation is of the general form $\exp(-\alpha y)$. This is the function which preserves its shape when differentiated, so we can expect the term in $y(d^2f/dy^2)$ to cancel against that in yf, and the term in df/dy to cancel against Ef. Explicitly, substituting the trial solution

$$f(y) = f_0 \exp(-\alpha y) \tag{6.93}$$

yields

$$-\frac{\hbar^2}{2m}[-2\alpha + 4y\alpha^2] + \tfrac{1}{2}Ky = E \tag{6.94}$$

after cancelling $f_0 \exp(-\alpha y)$ from both sides. In order for equation 6.94 to hold, the terms in y and the terms independent of y must vanish separately

$$-\frac{2\hbar^2}{m}\alpha^2 y + \tfrac{1}{2}Ky = 0 \tag{6.95}$$

$$\frac{\hbar^2}{m}\alpha = E \tag{6.96}$$

The first of these equations determines the range parameter α, and then the second gives the ground-state energy E. Explicitly

$$\alpha = \pm\frac{1}{2\hbar}(Km)^{1/2} \tag{6.97}$$

The solution with the minus sign is rejected since it gives $f(y) = \exp(\alpha y)$ and therefore $\phi(x) = \exp(\alpha x^2)$. This diverges at infinity, whereas the standing-wave condition is that ϕ should vanish at infinity. With the plus sign, equation 6.96 gives for the ground-state energy

$$E_0 = \tfrac{1}{2}\hbar\left(\frac{K}{m}\right)^{1/2} = \tfrac{1}{2}\hbar\omega_0 \tag{6.98}$$

where ω_0 is the frequency of the classical oscillator (equation 1.3). The ground-state wave function we have found is

$$\phi(x) = f_0 \exp[-(Km)^{1/2}x^2/2\hbar] \tag{6.99}$$

which indeed has the form sketched in figure 4.16. The amplitude f_0 does not come into the determination of the ground-state energy (it cancels from equation 6.94) and it is usually taken so that

$$\int_{-\infty}^{\infty} |\phi(x)|^2 \, dx = 1 \tag{6.100}$$

Since $|\phi(x)|^2$ is the probability that the particle is at the point x, equation 6.100 ensures that the overall probability that the particle is found somewhere is equal to unity.

We have now completed the programme of finding the ground-state, or zero-point, energy of the harmonic oscillator. A full solution would yield the heirarchy of allowed energy values, which is in fact

$$E_n = (n + \tfrac{1}{2})\hbar\omega_0 \tag{6.101}$$

where n is an integer. Equation 6.101 has the interesting and important consequence that the energy levels are equally spaced (figure 6.14).

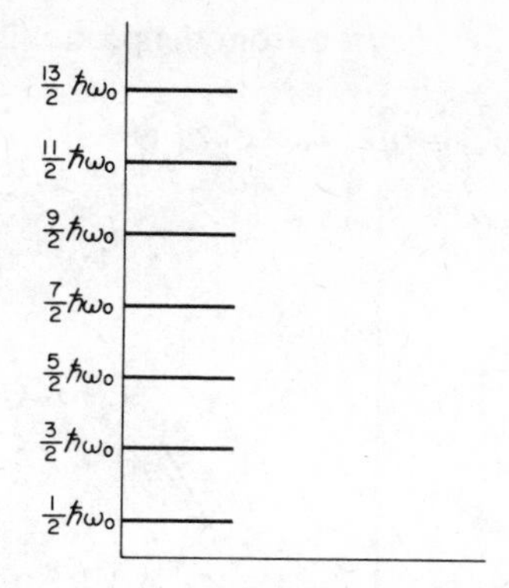

Figure 6.14 Energy-level scheme for the harmonic oscillator

6.9 The Hydrogen Atom

The potential energy of an electron in the neighbourhood of a proton is

$$V(r) = -\frac{e^2}{4\pi\epsilon_0 r} \tag{6.102}$$

where r is the distance of the electron from the proton. The potential is sketched in figure 6.15. In this problem, the electron moves in three dimensions, and has momenta p_x, p_y, p_z in the x, y and z directions. The equation for conservation of energy is therefore

$$\frac{p_x{}^2}{2m} + \frac{p_y{}^2}{2m} + \frac{p_z{}^2}{2m} - \frac{e^2}{4\pi\epsilon_0 r} = E \tag{6.103}$$

The wave function obviously depends upon x, y and z: $\psi = \psi(x, y, z)$. Application of the usual rule gives the time-independent Schrödinger equation

$$-\frac{\hbar^2}{2m}\left(\frac{\partial^2\psi}{\partial x^2} + \frac{\partial^2\psi}{\partial y^2} + \frac{\partial^2\psi}{\partial z^2}\right) - \frac{e^2}{4\pi\epsilon_0 r}\,\psi = E\psi \tag{6.104}$$

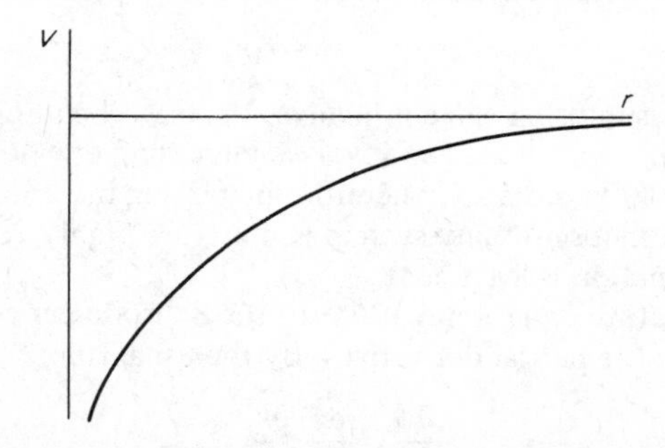

Figure 6.15 Potential function for the hydrogen atom

Note that the distance r of the electron from the proton is related to its co-ordinates (x, y, z) by

$$r = (x^2 + y^2 + z^2)^{1/2} \tag{6.105}$$

as can be seen from figure 6.16.

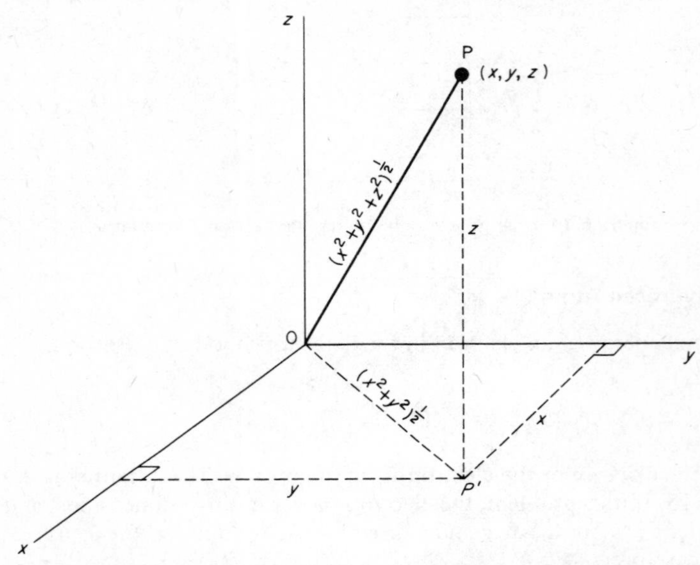

Figure 6.16 Electron at point P = (x, y, z). PP′ is the perpendicular on to the xy plane. The distance OP′ is $(x^2 + y^2)^{1/2}$ and PP′ is z, so the distance OP is $(x^2 + y^2 + z^2)^{1/2}$

Equation 6.104 is the Schrödinger equation from which the energy level scheme quoted in equation 4.22 can be derived. As for the harmonic oscillator, the general solution can be found by the series method. Once again, we shall simply calculate the ground-state energy, since this can be done in a more straightforward manner. We exploit the symmetry of the problem, and assume that the ground-state wave function depends on radial distance r only

$$\psi(x, y, z) = f(r) \tag{6.106}$$

This is a plausible assumption since it involves the least 'bending' of the wave function. Bending the wave function involves increasing one or more of the partial derivatives $\partial\psi/\partial x$, etc., and therefore increasing the kinetic energy T. The ground-state wave function of any system is therefore highly symmetrical since the symmetrical function is least bent.

In order to substitute equation 6.106 into the Schrödinger equation, we must find expressions for the partial derivatives. By the usual rule

$$\frac{\partial\psi}{\partial x} = \frac{\mathrm{d}f}{\mathrm{d}r}\frac{\partial r}{\partial x} \tag{6.107}$$

and from equation 6.105

$$\frac{\partial r}{\partial x} = \frac{x}{r} \tag{6.108}$$

so that

$$\frac{\partial \psi}{\partial x} = \frac{\mathrm{d}f}{\mathrm{d}r} \frac{x}{r} \tag{6.109}$$

The second derivative is

$$\begin{aligned}\frac{\partial^2 \psi}{\partial x^2} &= \frac{1}{r} \frac{\mathrm{d}f}{\mathrm{d}r} + x \frac{\mathrm{d}}{\mathrm{d}r}\left(\frac{1}{r} \frac{\mathrm{d}f}{\mathrm{d}r}\right) \frac{\partial r}{\partial x} \\ &= \frac{1}{r} \frac{\mathrm{d}f}{\mathrm{d}r} + \frac{x^2}{r}\left(-\frac{1}{r^2} \frac{\mathrm{d}f}{\mathrm{d}r} + \frac{1}{r} \frac{\mathrm{d}^2 f}{\mathrm{d}r^2}\right) \\ &= \left(\frac{1}{r} - \frac{x^2}{r^3}\right) \frac{\mathrm{d}f}{\mathrm{d}r} + \frac{x^2}{r^2} \frac{\mathrm{d}^2 f}{\mathrm{d}r^2}\end{aligned} \tag{6.110}$$

For the sum of the three partial derivatives, we find

$$\begin{aligned}\frac{\partial^2 \psi}{\partial x^2} + \frac{\partial^2 \psi}{\partial y^2} + \frac{\partial^2 \psi}{\partial z^2} &= \frac{3}{r} \frac{\mathrm{d}f}{\mathrm{d}r} - \frac{x^2 + y^2 + z^2}{r^3} \frac{\mathrm{d}f}{\mathrm{d}r} \\ &\quad + \frac{x^2 + y^2 + z^2}{r^2} \frac{\mathrm{d}^2 f}{\mathrm{d}r^2}\end{aligned} \tag{6.111}$$

Since $x^2 + y^2 + z^2 = r^2$, this simplifies to

$$\frac{\partial^2 \psi}{\partial x^2} + \frac{\partial^2 \psi}{\partial y^2} + \frac{\partial^2 \psi}{\partial z^2} = \frac{2}{r} \frac{\mathrm{d}f}{\mathrm{d}r} + \frac{\mathrm{d}^2 f}{\mathrm{d}r^2} \tag{6.112}$$

Substitution into equation 6.104 then yields

$$-\frac{\hbar^2}{2m}\left(\frac{\mathrm{d}^2 f}{\mathrm{d}r^2} + \frac{2}{r} \frac{\mathrm{d}f}{\mathrm{d}r}\right) - \frac{e^2}{4\pi\epsilon_0 r} f = Ef \tag{6.113}$$

This equation is rather similar to the one found for the harmonic oscillator, equation 6.92. For the same reason as given there, we substitute the trial solution

$$f = f_0 \exp(-\alpha r) \tag{6.114}$$

Equating to zero the terms in $r^{-1} \exp(-\alpha r)$ yields

$$\frac{\hbar^2}{m} \alpha - \frac{e^2}{4\pi\epsilon_0} = 0 \tag{6.115}$$

and from the terms in $\exp(-\alpha r)$

$$-\frac{\hbar^2\alpha^2}{2m} = E \tag{6.116}$$

As before, the first equation determines the range parameter α, and then the second gives the ground-state energy E

$$\alpha = \frac{e^2 m}{4\pi\epsilon_0\hbar^2} \tag{6.117}$$

$$E = -\frac{m}{2\hbar^2}\left(\frac{e^2}{4\pi\epsilon_0}\right)^2 \tag{6.118}$$

The ground-state energy is of course the same as quoted in equation 4.22, the latter being written in terms of $h = 2\pi\hbar$.

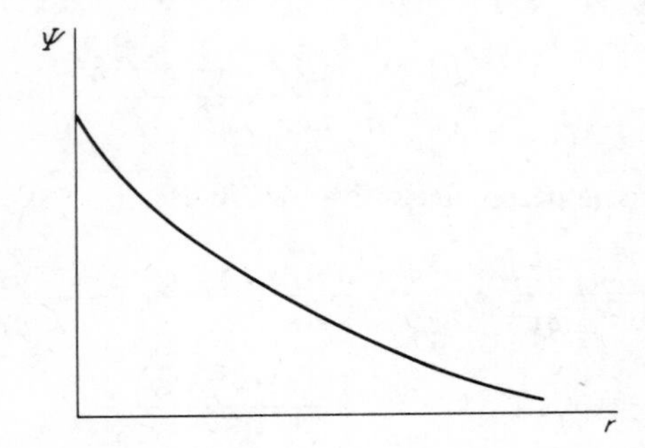

Figure 6.17 Ground-state wave function of the hydrogen atom as a function of radial distance r

The ground-state wave function of equation 6.114 decays as $r \to \infty$, as it must (figure 6.17). The range parameter α is determined by the balance between kinetic energy and potential energy. The higher energy levels (equation 4.22) are found by solving equation 6.104. The problem is a little more complicated than with the harmonic oscillator, because in general the dependence on all three co-ordinates x, y and z needs to be accounted for.

Problems

1. Confirm that equation 6.12 for the velocity v is dimensionally correct. (The dimensions of C can be found from equation 6.5.)

2. What are the dimensions of the specific acoustic impedance Z (equation 6.36)?

3. Equations 6.12 and 6.36 give the parameters v and Z in terms of C and ρ. Invert these equations to find C and ρ in terms of v and Z.

4. The following table gives values of Young's modulus, C, in units of 10^{10} N m^{-2}, and density ρ in units of 10^3 kg m^{-3} for various materials. Tabulate the corresponding values of sound velocity v and impedance Z for longitudinal waves on a bar.

	Al	Cu	Au	Fe	Crown glass	Quartz
C	7.05	13.1	7.8	21.2	6.5	7.3
ρ	2.7	8.9	19.3	7.9	2.9	2.6

5. Confirm that equation 6.57 for the conservation of energy at an interface does hold.

6. What should be the elastic constant C of an impedance matching layer of density 0.8×10^3 kg m^{-3} between copper ($C = 13.1 \times 10^{10}$ N m^{-2}, $\rho = 8.9 \times 10^3$ kg m^{-3}) and silver ($C = 10.9 \times 10^{10}$ N m^{-2}, $\rho = 10.5 \times 10^3$ kg m^{-3}). For waves of frequency 10^6 Hz, how thick is the matching layer? [UE]

7. Show that the interface transmission and reflection coefficients for the element velocity $\partial u/\partial t$ are

$$T_\text{v} = \frac{2}{1 + Z_2/Z_1}$$

$$R_\text{v} = \frac{Z_2 - Z_1}{Z_2 + Z_1}$$

that is, that they are identical in form to the electric field coefficients quoted in equations 6.63 and 6.64.

8. Derive the wave equation in one dimension for an elastic continuum of mass density ρ and elastic modulus C and show that the velocity of sound is equal to $(C/\rho)^{1/2}$. Discuss the applicability of this derivation to a one-dimensional lattice of atoms of mass m, distance a apart and with restoring force per unit displacement β. At what frequencies would you expect the derivation to be invalid? [UE]

9. A sinusoidal wave $u = u_0 \cos(kx - \omega t)$ travels in a one-dimensional elastic continuum (elastic constant C and mass per unit length ρ). Derive expressions for the following:

   (a) the particle velocity;
   (b) the average kinetic energy density;
   (c) the average potential energy density;
   (d) the average power;
   (e) the energy velocity. [UE]

10. An elastic wave in a metal impinges normally on the surface and some of it is

transmitted as a sound wave in air. The elastic constant and density of the metal are C_m and ρ_m, those of the air C_a and ρ_a.

(a) Define the transmission coefficient T;
(b) Derive an expression for T;
(c) Discuss how T is related to the transmitted intensity and estimate the fraction of incident intensity transmitted into the air using the following orders of magnitude:
$C_m = 10^{11}$ N m^{-2}, $\rho_m = 10^4$ kg m^{-3}, $C_a = 10^5$ N m^{-2}, $\rho_a = 1$ kg m^{-3}.
[UE]

11. Calculate the intensity (W m^{-2}) of light which would be required to exert a pressure, on a mirror from which it was totally reflected, equal to atmospheric pressure (10^5 N m^{-2}). What are the amplitudes of the electric and magnetic fields E and H in this light?

12. Consider a medium in which the elastic modulus C and the density ρ are functions of position $C(x)$ and $\rho(x)$. Show that the wave equation is

$$\frac{\partial}{\partial x}\left[C(x)\frac{\partial u}{\partial x}\right] = \rho(x)\frac{\partial^2 u}{\partial t^2}$$

13. The minimum intensity of light which can be detected by the eye is often quoted as five photons per second at a wavelength of 500 nm. Find the equivalent intensity of a continuous wave, and the amplitudes E_0 and H_0 in the continuous wave. What is the force exerted on the eye by the light? (Assume diameter of eye = 0.5 cm.)

14. With the assumptions of worked example 6.1 (Z_1 fixed and Z_2 variable), sketch the reflected and transmitted intensities Z_1R^2 and Z_2T^2 as functions of Z_2. See that your sketches agree with equation 6.57 for conservation of energy.

Appendixes

Appendix I Useful Trigonometric Identities

Collected here for convenience are some identities which are used at various places in the book.

A1.1 Sine and cosine functions (sinusoidal functions)

Graphs are shown in figure A1.1.

$$\sin\left(x + \frac{\pi}{2}\right) = \cos x$$

$$\sin(x + \pi) = -\sin x$$

$$\sin\left(x + \frac{3\pi}{2}\right) = -\cos x$$

$$\sin(x + 2\pi) = \sin x$$

$$\sin^2 x + \cos^2 x = 1$$

A1.2 Addition and composition formulae

$$\sin(A \pm B) = \sin A \cos B \pm \cos A \sin B$$

$$\cos(A \pm B) = \cos A \cos B \mp \sin A \sin B$$

$$\sin 2A = 2 \sin A \cos A$$

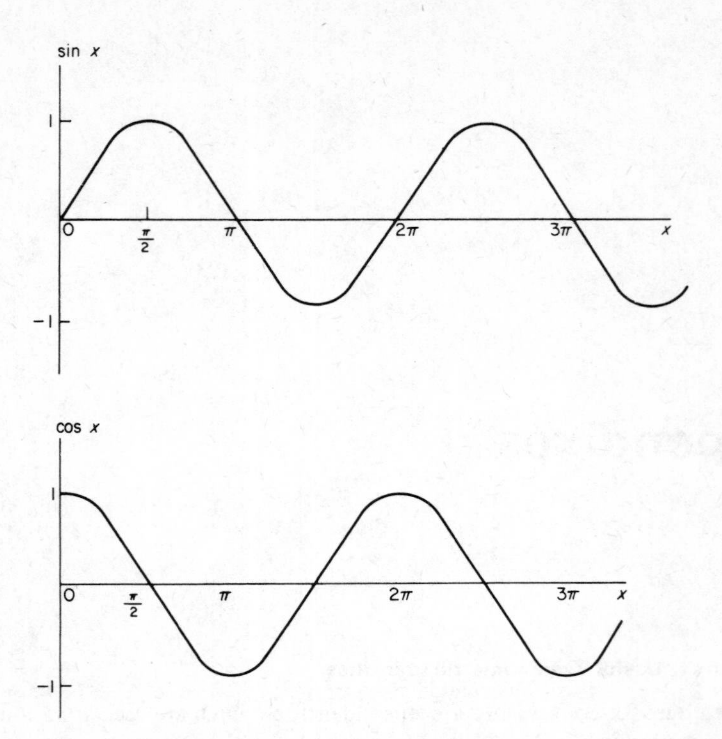

Figure A1.1 Sin x and cos x as functions of x

$$\cos 2A = \cos^2 A - \sin^2 A = 2\cos^2 A - 1 = 1 - 2\sin^2 A$$

$$\sin A + \sin B = 2 \sin \tfrac{1}{2}(A + B) \cos \tfrac{1}{2}(A - B)$$

$$\sin A - \sin B = 2 \cos \tfrac{1}{2}(A + B) \sin \tfrac{1}{2}(A - B)$$

$$\cos A + \cos B = 2 \cos \tfrac{1}{2}(A + B) \cos \tfrac{1}{2}(A - B)$$

$$\cos A - \cos B = 2 \sin \tfrac{1}{2}(A + B) \sin \tfrac{1}{2}(B - A)$$

A1.3 Relation to exponential function

$$\exp(i\theta) = \cos\theta + i \sin\theta$$

$$\exp(-i\theta) = \cos\theta - i \sin\theta$$

$$\cos\theta = \tfrac{1}{2}[\exp(i\theta) + \exp(-i\theta)]$$

$$\sin\theta = \tfrac{1}{2}[\exp(i\theta) - \exp(-i\theta)]$$

It is often convenient to represent a sinusoidal function as the real part of an exponential function. We write

$$A_0 \cos(\omega t + \delta) = \text{Re}\, A \exp(i\omega t)$$

where the amplitude A is complex

$$A = A_0 \exp(i\delta)$$

so that A contains both the real amplitude A_0 and the phase δ. It is common to leave out 'Re' in front of the complex expression, as in some places in the book, on the understanding that the real part is to be taken.

This convention needs handling with care when it is necessary to multiply two sinusoidal quantities, for example voltage and current in an a.c. circuit. For two general complex numbers z_1 and z_2

$$\text{Re}\,(z_1 z_2) \neq \text{Re}\, z_1 \; \text{Re}\, z_2$$

as can be verified by expressing both z_1 and z_2 as real plus imaginary parts. This means that we must take the real part explicitly before multiplying sinusoidal functions.

A1.4 Time averaging

We frequently want the time average of $\cos^2(\omega t + \delta)$ which is defined as

$$\langle \cos^2(\omega t + \delta) \rangle = T^{-1} \int_0^T \cos^2(\omega t + \delta)\, dt$$

where T is any repeat time (periodic time) of $\cos^2(\omega t + \delta)$. We may take $T = 2\pi/\omega$, so that

$$\langle \cos^2(\omega t + \delta) \rangle = \frac{\omega}{2\pi} \int_0^{2\pi/\omega} \cos^2(\omega t + \delta)\, dt$$

We write

$$\cos^2(\omega t + \delta) = \tfrac{1}{2}[1 + \cos(2\omega t + 2\delta)]$$

so that

$$\langle \cos^2(\omega t + \delta) \rangle = \frac{\omega}{2\pi} \int_0^{2\pi/\omega} \tfrac{1}{2}\, dt + \frac{\omega}{2\pi} \int_0^{2\pi/\omega} \cos(2\omega t + 2\delta)\, dt$$

or finally

$$\langle \cos^2(\omega t + \delta) \rangle = \tfrac{1}{2}$$

since the second integral is zero.

An important application of this formula is to calculating the intensity of an electromagnetic wave. The *amplitude* of the wave is the magnitude of an E field, in $\text{V}\,\text{m}^{-1}$, which may be written as

$$u = \text{Re}\, u_0 \exp(i\omega t)$$

The intensity I is measured in W m^{-2}, and just like the power in an electrical circuit it is proportional to the mean square of the voltage level

$$I = \frac{1}{Z} \langle u^2 \rangle$$

It should be noted that the constant Z has the dimension of ohms. Its magnitude depends on the medium through which the wave is travelling. Substituting in the expression for u we find

$$I = \frac{1}{2Z} (\mathrm{Re}\, u_0)^2$$

Appendix 2 Some Fundamental Constants

c	velocity of light	3×10^8 m s^{-1}
e	electronic charge	1.60×10^{-19} C
h	Planck's constant	6.58×10^{-34} J s (4.14×10^{-15} eV s)
$\hbar$	Planck's constant/2π	1.06×10^{-34} J s
m_o	electron rest mass	9.11×10^{-31} kg
m_p	proton rest mass	1.67×10^{-27} kg

The following table gives the SI submultiples and multiples of basic units:

Fraction	Prefix	Symbol	Multiple	Prefix	Symbol
10^{-1}	deci	d	10	deka	da
10^{-2}	centi	c	10^2	hecto	h
10^{-3}	milli	m	10^3	kilo	k
10^{-6}	micro	μ	10^6	mega	M
10^{-9}	nano	n	10^9	giga	G
10^{-12}	pico	p	10^{12}	tera	T
10^{-15}	femto	f			
10^{-18}	atto	a			

Index